Learn, Practice, Succeed

Eureka Math®
Grade 8
Module 2

Published by Great Minds®.

Copyright © 2019 Great Minds®.

Printed in the U.S.A.

This book may be purchased from the publisher at eureka-math.org.

10 9 8 7 6 5 4 3 2 1

ISBN 978-1-64054-981-4

G8-M2-LPS-05.2019

Students, families, and educators:

Thank you for being part of the *Eureka Math*® community, where we celebrate the joy, wonder, and thrill of mathematics.

In *Eureka Math* classrooms, learning is activated through rich experiences and dialogue. That new knowledge is best retained when it is reinforced with intentional practice. The *Learn, Practice, Succeed* book puts in students' hands the problem sets and fluency exercises they need to express and consolidate their classroom learning and master grade-level mathematics. Once students learn and practice, they know they can succeed.

What is in the Learn, Practice, Succeed *book?*

Fluency Practice: Our printed fluency activities utilize the format we call a Sprint. Instead of rote recall, Sprints use patterns across a sequence of problems to engage students in reasoning and to reinforce number sense while building speed and accuracy. Sprints are inherently differentiated, with problems building from simple to complex. The tempo of the Sprint provides a low-stakes adrenaline boost that increases memory and automaticity.

Classwork: A carefully sequenced set of examples, exercises, and reflection questions support students' in-class experiences and dialogue. Having classwork preprinted makes efficient use of class time and provides a written record that students can refer to later.

Exit Tickets: Students show teachers what they know through their work on the daily Exit Ticket. This check for understanding provides teachers with valuable real-time evidence of the efficacy of that day's instruction, giving critical insight into where to focus next.

Homework Helpers and Problem Sets: The daily Problem Set gives students additional and varied practice and can be used as differentiated practice or homework. A set of worked examples, Homework Helpers, support students' work on the Problem Set by illustrating the modeling and reasoning the curriculum uses to build understanding of the concepts the lesson addresses.

Homework Helpers and Problem Sets from prior grades or modules can be leveraged to build foundational skills. When coupled with *Affirm*®, *Eureka Math*'s digital assessment system, these Problem Sets enable educators to give targeted practice and to assess student progress. Alignment with the mathematical models and language used across *Eureka Math* ensures that students notice the connections and relevance to their daily instruction, whether they are working on foundational skills or getting extra practice on the current topic.

Where can I learn more about Eureka Math *resources?*

The Great Minds® team is committed to supporting students, families, and educators with an ever-growing library of resources, available at eureka-math.org. The website also offers inspiring stories of success in the *Eureka Math* community. Share your insights and accomplishments with fellow users by becoming a *Eureka Math* Champion.

Best wishes for a year filled with "aha" moments!

Jill Diniz

Jill Diniz
Chief Academic Officer, Mathematics
Great Minds

Contents

Module 2: The Concept of Congruence

Exploratory Challenge

a.   Describe, intuitively, what kind of transformation is required to move the figure on the left to each of the figures (1)–(3) on the right. To help with this exercise, use a transparency to copy the figure on the left. Note: Begin by moving the left figure to each of the locations in (1), (2), and (3).

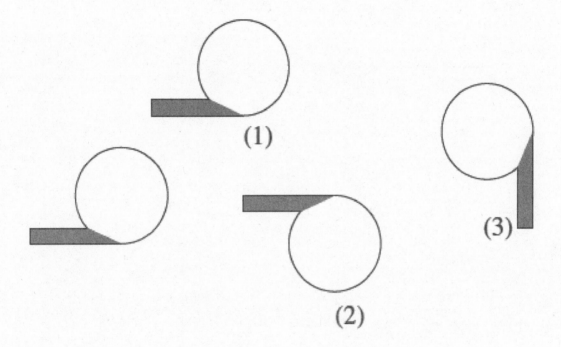

b. Given two segments AB and CD, which could be very far apart, how can we find out if they have the same length without measuring them individually? Do you think they have the same length? How do you check? In other words, why do you think we need to move things around on the plane?

© 2019 Great Minds®. eureka-math.org

EUREKA MATH®

Lesson Summary

A *transformation F* of the plane is a function that assigns to each point P of the plane a point $F(P)$ in the plane.

- By definition, the symbol $F(P)$ denotes a specific single point, unambiguously.
- The point $F(P)$ will be called the image of P by F. Sometimes the image of P by F is denoted simply as P' (read "P prime").
- The transformation F is sometimes said to "move" the point P to the point $F(P)$.
- We also say F maps P to $F(P)$.

In this module, we will mostly be interested in transformations that are given by rules, that is, a set of step-by-step instructions that can be applied to any point P in the plane to get its image.

If given any two points P and Q, the distance between the images $F(P)$ and $F(Q)$ is the same as the distance between the original points P and Q, and then the transformation F preserves distance, or is distance-preserving.

- A distance-preserving transformation is called a *rigid motion* (or an *isometry*), and the name suggests that it moves the points of the plane around in a rigid fashion.

Name _____   Date _____

First, draw a simple figure and name it Figure W. Next, draw its image under some transformation (i.e., trace your Figure W on the transparency), and then move it. Finally, draw its image somewhere else on the paper.

Describe, intuitively, how you moved the figure. Use complete sentences.

Lesson Notes

Transformations of the plane (i.e., translations, reflections, and rotations) are introduced. Transformations are distance preserving.

Example

Using as much of the new vocabulary as you can, try to describe what you see in the diagram below.

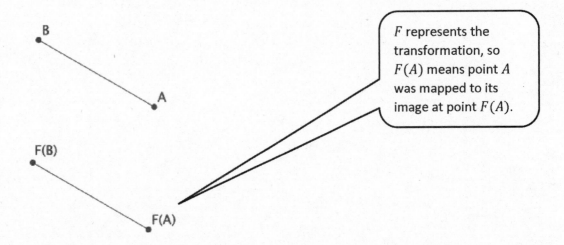

F represents the transformation, so $F(A)$ means point A was mapped to its image at point $F(A)$.

*There was a transformation, **F**, that moved point A to its image $F(A)$ and point B to its image $F(B)$. Since a transformation preserves distance, the distance between points A and B is the same as the distance between points $F(A)$ and $F(B)$.*

1. Using as much of the new vocabulary as you can, try to describe what you see in the diagram below.

2. Describe, intuitively, what kind of transformation is required to move Figure A on the left to its image on the right.

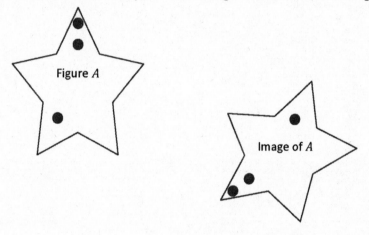

Exercise 1

Draw at least three different vectors, and show what a translation of the plane along each vector looks like. Describe what happens to the following figures under each translation using appropriate vocabulary and notation as needed.

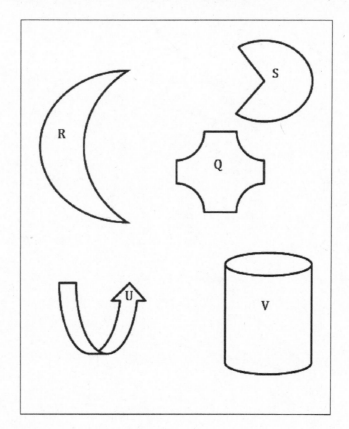

The diagram below shows figures and their images under a translation along $\overrightarrow{HI}$. Use the original figures and the translated images to fill in missing labels for points and measures.

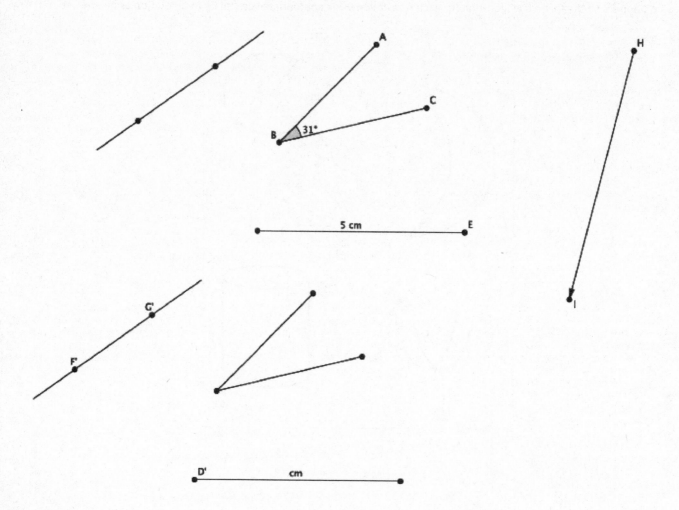

Lesson Summary

Translation occurs along a given vector:

- A *vector* is directed line segment, that is, it is a segment with a direction given by connecting one of its endpoint (called the *initial point* or *starting point*) to the other endpoint (called the *terminal point* or simply the *endpoint*). It is often represented as an "arrow" with a "tail" and a "tip."

- The *length of a vector* is, by definition, the length of its underlying segment.

- Pictorially note the starting and endpoints:

A translation of a plane along a given vector is a basic rigid motion of a plane.

The three basic properties of translation are as follows:

(Translation 1) A translation maps a line to a line, a ray to a ray, a segment to a segment, and an angle to an angle.

(Translation 2) A translation preserves lengths of segments.

(Translation 3) A translation preserves measures of angles.

Terminology

TRANSLATION (description): For vector $\overrightarrow{AB}$, a *translation along* $\overrightarrow{AB}$ is the transformation of the plane that maps each point C of the plane to its image C' so that the line $\overleftrightarrow{CC'}$ is parallel to the vector (or contains it), and the vector $\overrightarrow{CC'}$ points in the same direction and is the same length as the vector $\overrightarrow{AB}$.

Name _____   Date _____

1. Name the vector in the picture below.

2. Name the vector along which a translation of a plane would map point A to its image $T(A)$.

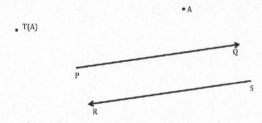

3. Is Maria correct when she says that there is a translation along a vector that maps segment AB to segment CD? If so, draw the vector. If not, explain why not.

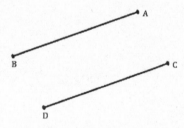

4. Assume there is a translation that maps segment AB to segment CD shown above. If the length of segment CD is 8 units, what is the length of segment AB? How do you know?

Lesson Notes

Translations move figures along a vector. The vector has a starting point and an endpoint. Translations map lines to lines, rays to rays, segments to segments, and angles to angles. Translations preserve lengths of segments and degrees of angles.

Examples

1. Use your transparency to translate the angle of 32 degrees, a segment with length 1.5 in., a point, and a circle with radius 2 cm along vector $\overrightarrow{AB}$. Label points and measures (measurements do not need to be precise, but your figure must be labeled correctly). Sketch the images of the translated figures, and label them.

    Note: The figures below have not been drawn to scale.

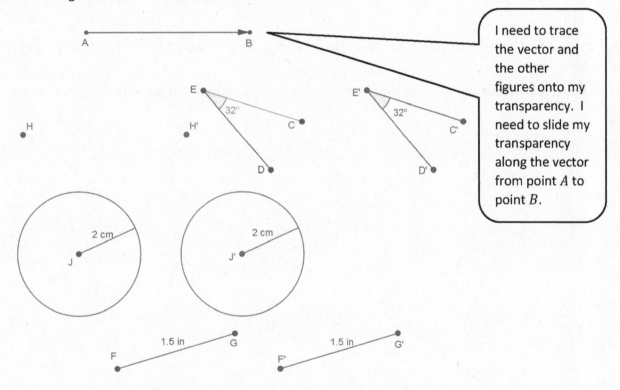

I need to trace the vector and the other figures onto my transparency. I need to slide my transparency along the vector from point A to point B.

Use your drawing from Problem 1 to answer the questions below.

2. What is the length of the translated segment? How does this length compare to the length of the original segment? Explain.

   1.5 inches. The length is the same as the original because translations preserve the lengths of segments.

3. What is the length of the radius in the translated circle? How does this radius length compare to the radius of the original circle? Explain.

   2 centimeters. The length is the same as the original because translations preserve the lengths of segments.

4. What is the degree of the translated angle? How does this degree compare to the degree of the original angle? Explain.

   32 degrees. The angles will have the same measure because translations preserve degrees of angles.

1. Translate the plane containing Figure A along $\overrightarrow{AB}$. Use your transparency to sketch the image of Figure A by this translation. Mark points on Figure A, and label the image of Figure A accordingly.

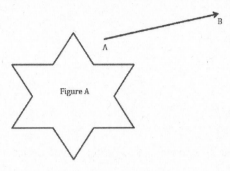

2. Translate the plane containing Figure B along $\overrightarrow{BA}$. Use your transparency to sketch the image of Figure B by this translation. Mark points on Figure B, and label the image of Figure B accordingly.

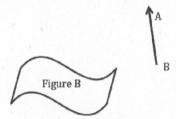

3. Draw an acute angle (your choice of degree), a segment with length 3 cm, a point, a circle with radius 1 in., and a vector (your choice of length, i.e., starting point and ending point). Label points and measures (measurements do not need to be precise, but your figure must be labeled correctly). Use your transparency to translate all of the figures you have drawn along the vector. Sketch the images of the translated figures and label them.

4. What is the length of the translated segment? How does this length compare to the length of the original segment? Explain.

5. What is the length of the radius in the translated circle? How does this radius length compare to the radius of the original circle? Explain.

6. What is the degree of the translated angle? How does this degree compare to the degree of the original angle? Explain.

7. Translate point D along vector $\overrightarrow{AB}$, and label the image D'. What do you notice about the line containing vector $\overrightarrow{AB}$ and the line containing points D and D'? (Hint: Will the lines ever intersect?)

D.

A   B

8. Translate point E along vector $\overrightarrow{AB}$, and label the image E'. What do you notice about the line containing vector $\overrightarrow{AB}$ and the line containing points E and E'?

B

E

A

EUREKA MATH

1. Draw a line passing through point P that is parallel to line L. Draw a second line passing through point P that is parallel to line L and that is distinct (i.e., different) from the first one. What do you notice?

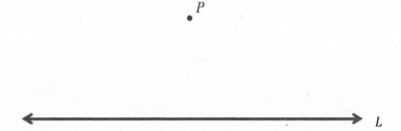

2. Translate line L along the vector $\overrightarrow{AB}$. What do you notice about L and its image, L'?

3. Line L is parallel to vector $\overrightarrow{AB}$. Translate line L along vector $\overrightarrow{AB}$. What do you notice about L and its image, L'?

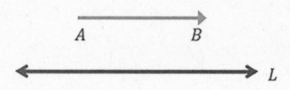

4. Translate line L along the vector $\overrightarrow{AB}$. What do you notice about L and its image, L'?

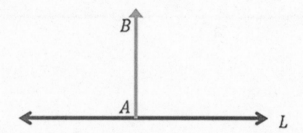

EUREKA MATH

5. Line L has been translated along vector $\overrightarrow{AB}$, resulting in L'. What do you know about lines L and L'?

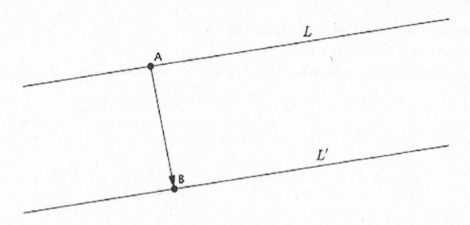

6. Translate L_1 and L_2 along vector $\overrightarrow{DE}$. Label the images of the lines. If lines L_1 and L_2 are parallel, what do you know about their translated images?

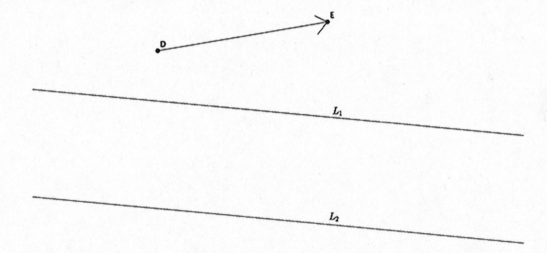

Lesson Summary

- Two lines in the plane are parallel if they do not intersect.

- Translations map parallel lines to parallel lines.

- Given a line L and a point P not lying on L, there is at most one line passing through P and parallel to L.

EUREKA
MATH®

Name _____   Date _____

1. Translate point Z along vector $\overrightarrow{AB}$. What do you know about the line containing vector $\overrightarrow{AB}$ and the line formed when you connect Z to its image Z'?

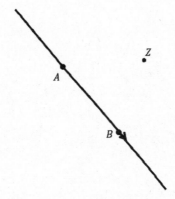

2. Using the above diagram, what do you know about the lengths of segments ZZ' and AB?

3. Let points A and B be on line L and the vector $\overrightarrow{AC}$ be given, as shown below. Translate line L along vector $\overrightarrow{AC}$. What do you know about line L and its image, L'? How many other lines can you draw through point C that have the same relationship as L and L'? How do you know?

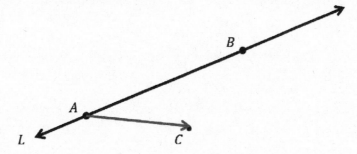

1. Translate ∠XYZ, point A, point B, and rectangle HIJK along vector $\overrightarrow{EF}$. Sketch the images, and label all points using prime notation.

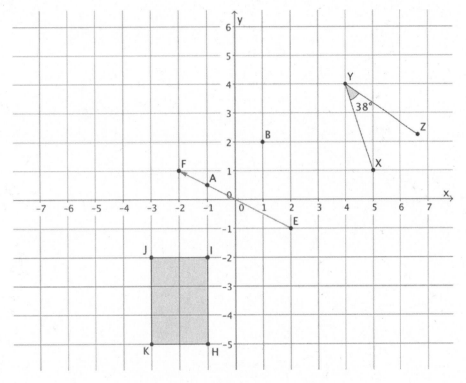

2. What is the measure of the translated image of ∠XYZ? How do you know?

3. Connect B to B'. What do you know about the line that contains the segment formed by BB' and the line containing the vector $\overrightarrow{EF}$?

4. Connect A to A'. What do you know about the line that contains the segment formed by AA' and the line containing the vector $\overrightarrow{EF}$?

5. Given that figure HIJK is a rectangle, what do you know about lines that contain segments HI and JK and their translated images? Explain.

1.   Reflect △ *ABC* and Figure *D* across line *L*. Label the reflected images.

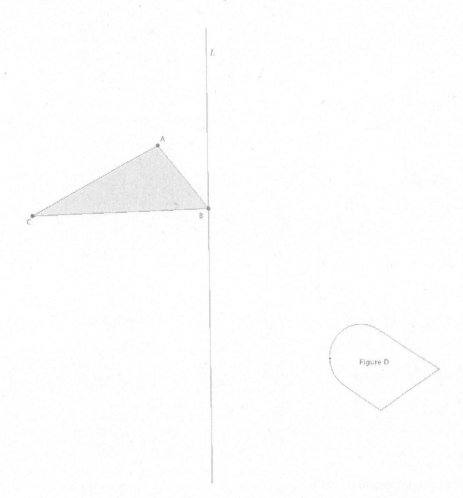

2.   Which figure(s) were not moved to a new location on the plane under this transformation?

3. Reflect the images across line L. Label the reflected images.

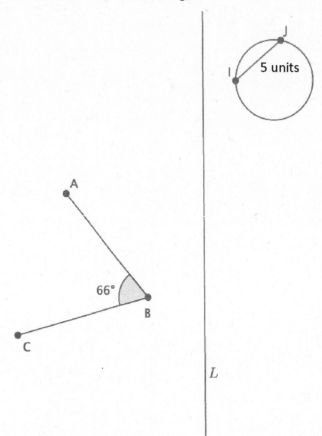

4. Answer the questions about the image above.

    a.   Use a protractor to measure the reflected $\angle ABC$. What do you notice?

    b.   Use a ruler to measure the length of IJ and the length of the image of IJ after the reflection. What do you notice?

EUREKA
MATH

5. Reflect Figure R and $\triangle EFG$ across line L. Label the reflected images.

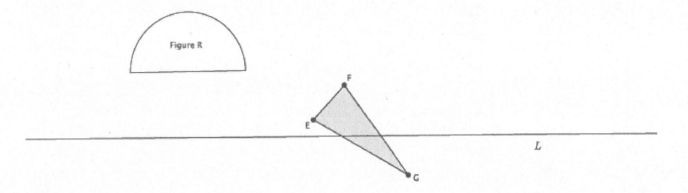

Basic Properties of Reflections:

(Reflection 1) A reflection maps a line to a line, a ray to a ray, a segment to a segment, and an angle to an angle.

(Reflection 2) A reflection preserves lengths of segments.

(Reflection 3) A reflection preserves measures of angles.

If the reflection is across a line L and P is a point not on L, then L bisects and is perpendicular to the segment PP', joining P to its reflected image P'. That is, the lengths of OP and OP' are equal.

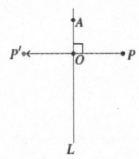

Use the picture below for Exercises 6–9.

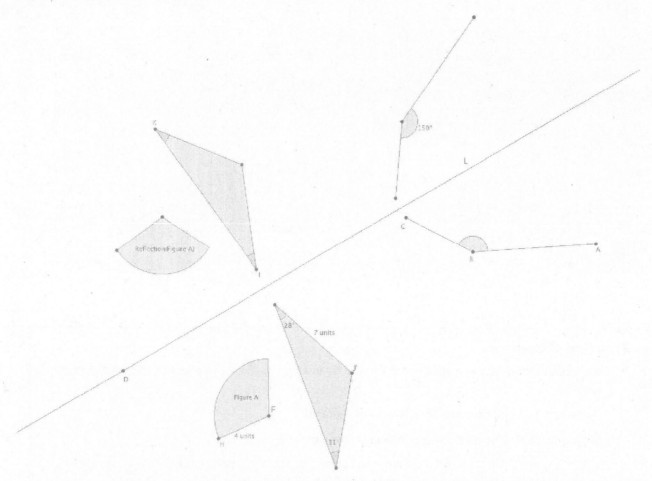

6. Use the picture to label the unnamed points.

7. What is the measure of ∠*JKI*? ∠*KIJ*? ∠*ABC*? How do you know?

8. What is the length of segment *Reflection*(*FH*)? *IJ*? How do you know?

9. What is the location of *Reflection*(*D*)? Explain.

EUREKA
MATH®

Lesson Summary

- A reflection is another type of basic rigid motion.

- A reflection across a line maps one half-plane to the other half-plane; that is, it maps points from one side of the line to the other side of the line. The reflection maps each point on the line to itself. The line being reflected across is called the *line of reflection*.

- When a point P is joined with its reflection P' to form the segment PP', the line of reflection bisects and is perpendicular to the segment PP'.

Terminology

REFLECTION (description): Given a line L in the plane, a *reflection across* L is the transformation of the plane that maps each point on the line L to itself, and maps each remaining point P of the plane to its image P' such that L is the perpendicular bisector of the segment PP'.

Name _____   Date _____

1. Let there be a reflection across line L_{AB}. Reflect $\Delta\, CDE$ across line L_{AB}. Label the reflected image.

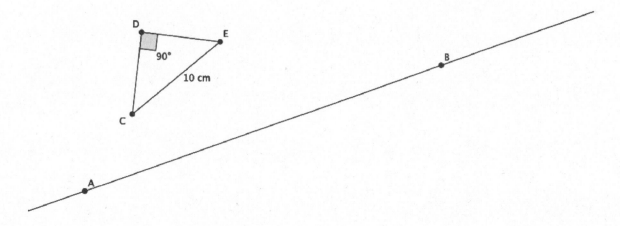

Picture not drawn to scale.

2. Use the diagram above to state the measure of *Reflection*($\angle CDE$). Explain.

3. Use the diagram above to state the length of segment *Reflection*(CE). Explain.

4. Connect point C to its image in the diagram above. What is the relationship between line L_{AB} and the segment that connects point C to its image?

Reflections are a basic rigid motion that maps lines to lines, rays to rays, segments to segments, and angles to angles. Basic rigid motions preserve lengths of segments and degrees of measures of angles. Reflections occur across a line called the line of reflection.

Examples

1. In the diagram below, $\angle ABC = 112°$, $AC = 6.3$ cm, $EF = 0.8$ cm, point H is on line L, and point G is off of line L. Let there be a reflection across line L. Reflect and label each of the figures, and answer the questions that follow.

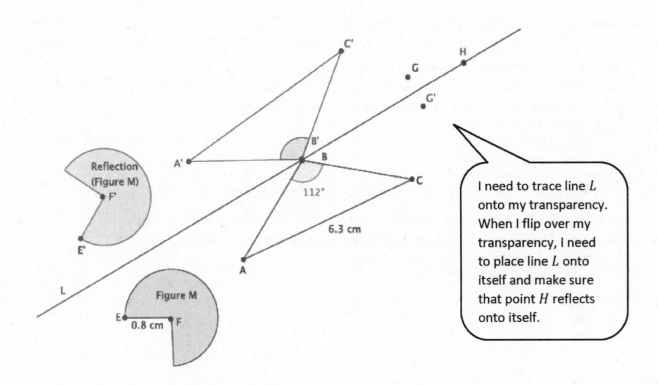

I need to trace line L onto my transparency. When I flip over my transparency, I need to place line L onto itself and make sure that point H reflects onto itself.

Note: Diagram not to scale.

2. What is the measure of *Reflection*(∠*ABC*)? Explain.

   The measure of Reflection(∠ABC) is 112°. Reflections preserve degrees of angles.

3. What is the length of *Reflection*(*EF*)? Explain.

   The length of Reflection(EF) is 0.8 cm. Reflections preserve lengths of segments.

4. What is the length of *Reflection*(*AC*)?

   The length of Reflection(AC) is 6.3 cm.

5. Three figures in the picture were not moved under the reflection. Name the three figures, and explain why they were not moved.

   Point B, point H, and line L were not moved. All of the points that make up the line of reflection remain in the same location when reflected. Since points B and H are on the line of reflection, they were not moved.

   > I remember my teacher telling me that the line of reflection is reflected, but it isn't moved to a new location.

6. Connect points *G* and *G′*. Name the point of intersection of the segment with the line of reflection point *Q*. What do you know about the lengths of segments *QG* and *QG′*?

   Segments QG and QG′ are equal in length. The segment GG′ connects point G to its image, G′. The line of reflection will go through the midpoint of, or bisect, the segment created when you connect a point to its image.

Lesson 4:   Definition of Reflection and Basic Properties

EUREKA MATH

1. In the picture below, $\angle DEF = 56°$, $\angle ACB = 114°$, $AB = 12.6$ units, $JK = 5.32$ units, point E is on line L, and point I is off of line L. Let there be a reflection across line L. Reflect and label each of the figures, and answer the questions that follow.

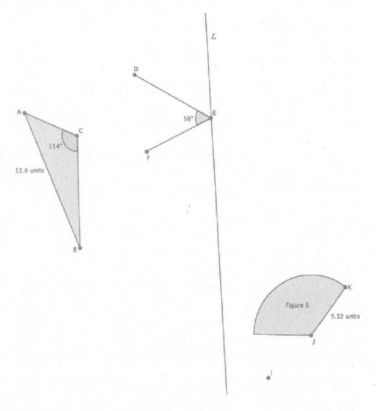

2. What is the measure of $Reflection(\angle DEF)$? Explain.

3. What is the length of $Reflection(JK)$? Explain.

4. What is the measure of $Reflection(\angle ACB)$?

5. What is the length of $Reflection(AB)$?

6. Two figures in the picture were not moved under the reflection. Name the two figures, and explain why they were not moved.

7. Connect points I and I'. Name the point of intersection of the segment with the line of reflection point Q. What do you know about the lengths of segments IQ and QI'?

Exercises

1. Let there be a rotation of d degrees around center O. Let P be a point other than O. Select d so that $d \geq 0$. Find P' (i.e., the rotation of point P) using a transparency.

2. Let there be a rotation of d degrees around center O. Let P be a point other than O. Select d so that $d < 0$. Find P' (i.e., the rotation of point P) using a transparency.

3. Which direction did the point P rotate when $d \geq 0$?

4. Which direction did the point P rotate when $d < 0$?

5. Let L be a line, $\overrightarrow{AB}$ be a ray, $\overline{CD}$ be a segment, and $\angle EFG$ be an angle, as shown. Let there be a rotation of d degrees around point O. Find the images of all figures when $d \geq 0$.

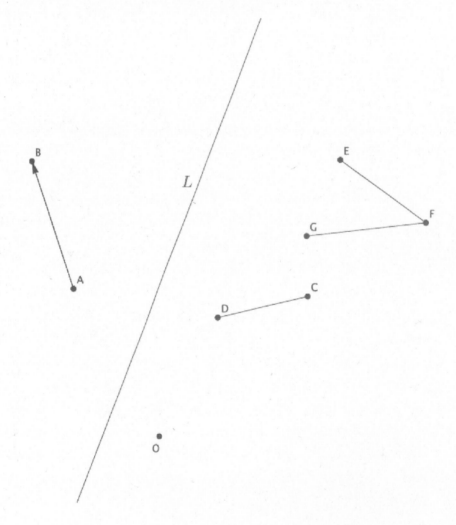

6. Let $\overline{AB}$ be a segment of length 4 units and $\angle CDE$ be an angle of size 45°. Let there be a rotation by d degrees, where $d < 0$, about O. Find the images of the given figures. Answer the questions that follow.

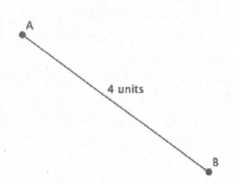

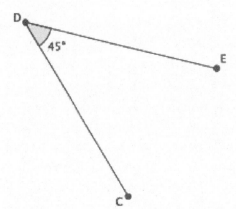

a. What is the length of the rotated segment $Rotation(AB)$?

b. What is the degree of the rotated angle $Rotation(\angle CDE)$?

7. Let L_1 and L_2 be parallel lines. Let there be a rotation by d degrees, where $-360 < d < 360$, about O. Is $(L_1)' \parallel (L_2)'$?

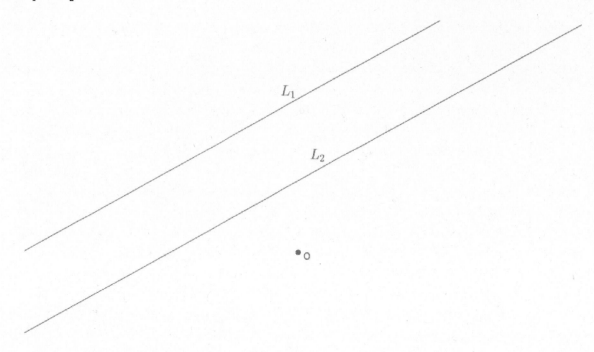

8. Let L be a line and O be the center of rotation. Let there be a rotation by d degrees, where $d \neq 180$ about O. Are the lines L and L' parallel?

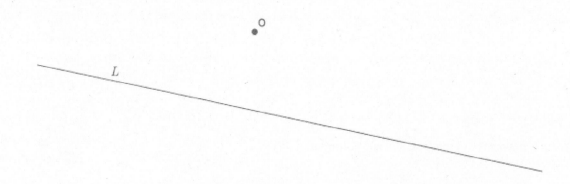

Lesson 5:   Definition of Rotation and Basic Properties

EUREKA MATH

Lesson Summary

Rotations require information about the center of rotation and the degree in which to rotate. Positive degrees of rotation move the figure in a counterclockwise direction. Negative degrees of rotation move the figure in a clockwise direction.

Basic Properties of Rotations:

- (Rotation 1) A rotation maps a line to a line, a ray to a ray, a segment to a segment, and an angle to an angle.
- (Rotation 2) A rotation preserves lengths of segments.
- (Rotation 3) A rotation preserves measures of angles.

When parallel lines are rotated, their images are also parallel. A line is only parallel to itself when rotated exactly 180°.

Terminology

ROTATION (DESCRIPTION): For a number d between 0 and 180, the *rotation of d degrees around center O* is the transformation of the plane that maps the point O to itself, and maps each remaining point P of the plane to its image P' in the counterclockwise half-plane of ray $\overrightarrow{OP}$ so that P and P' are the same distance away from O and the measurement of $\angle P'OP$ is d degrees.

The *counterclockwise half-plane* is the half-plane that lies to the left of $\overrightarrow{OP}$ while moving along $\overrightarrow{OP}$ in the direction from O to P.

Name _____   Date _____

1. Given the figure H, let there be a rotation by d degrees, where $d \geq 0$, about O. Let $Rotation(H)$ be H'. Note the direction of the rotation with an arrow.

2. Using the drawing above, let $Rotation_1$ be the rotation d degrees with $d < 0$, about O. Let $Rotation_1(H)$ be H''. Note the direction of the rotation with an arrow.

Lesson Notes

Rotations are a basic rigid motion that maps lines to lines, rays to rays, segments to segments, and angles to angles. Rotations preserve lengths of segments and degrees of measures of angles. Rotations require information about the center of rotation and the degree in which to rotate. Positive degrees of rotation move the figure in a counterclockwise direction. Negative degrees of rotation move the figure in a clockwise direction.

Examples

1. Let there be a rotation by 90° around the center O.

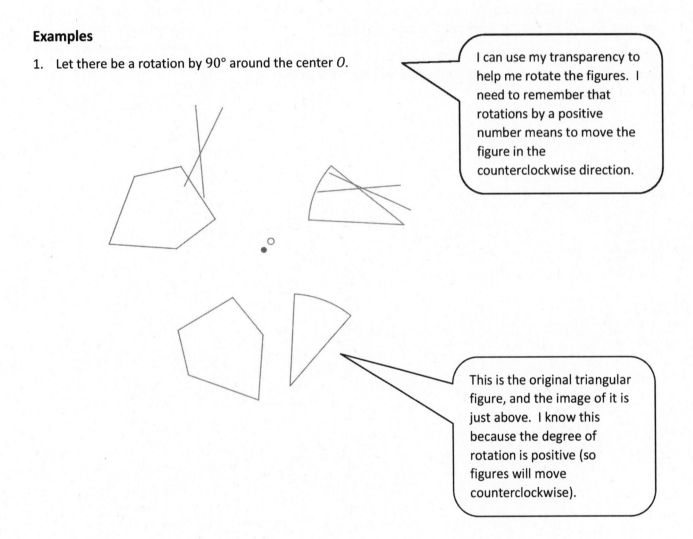

I can use my transparency to help me rotate the figures. I need to remember that rotations by a positive number means to move the figure in the counterclockwise direction.

This is the original triangular figure, and the image of it is just above. I know this because the degree of rotation is positive (so figures will move counterclockwise).

2. A segment of length 18 in. has been rotated d degrees around a center O. What is the length of the rotated segment? How do you know?

   The rotated segment will be 18 in. in length. (Rotation 2) states that rotations preserve lengths of segments, so the length of the rotated segment will remain the same as the original.

3. An angle of size 52° has been rotated d degrees around a center O. What is the size of the rotated angle? How do you know?

   The rotated angle will be 52°. (Rotation 3) states that rotations preserve the measures of angles, so the rotated angle will be the same size as the original.

> I need to remember that it doesn't matter how many degrees I rotate, the basic properties will be true. I can find the numbered Basic Properties of Rotation in my Lesson Summary box.

EUREKA
MATH®

1. Let there be a rotation by −90° around the center O.

• O

2. Explain why a rotation of 90 degrees around any point O never maps a line to a line parallel to itself.

3. A segment of length 94 cm has been rotated d degrees around a center O. What is the length of the rotated segment? How do you know?

4. An angle of size 124° has been rotated d degrees around a center O. What is the size of the rotated angle? How do you know?

Example 1

The picture below shows what happens when there is a rotation of 180° around center O.

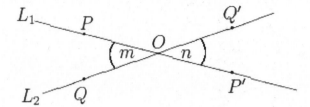

Example 2

The picture below shows what happens when there is a rotation of 180° around center O, the origin of the coordinate plane.

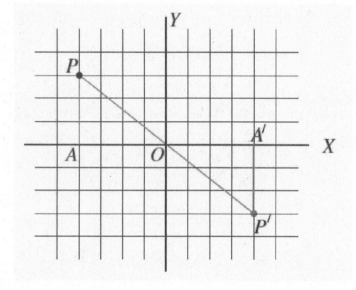

1. Using your transparency, rotate the plane 180 degrees, about the origin. Let this rotation be $Rotation_0$. What are the coordinates of $Rotation_0 (2, -4)$?

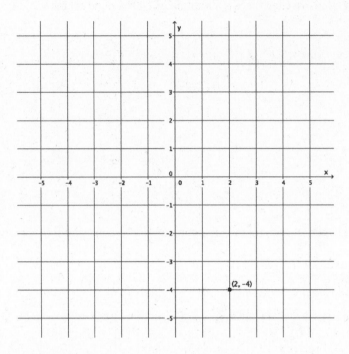

2. Let $Rotation_0$ be the rotation of the plane by 180 degrees, about the origin. <u>Without</u> using your transparency, find $Rotation_0(-3, 5)$.

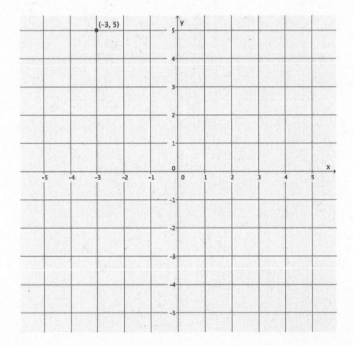

EUREKA
MATH®

3. Let $Rotation_0$ be the rotation of 180 degrees around the origin. Let L be the line passing through $(-6, 6)$ parallel to the x-axis. Find $Rotation_0(L)$. Use your transparency if needed.

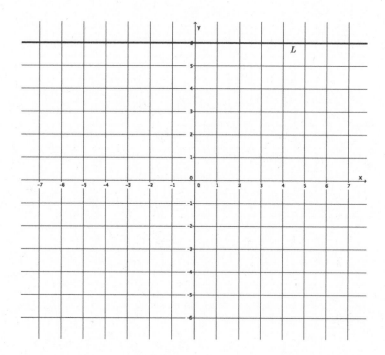

4. Let $Rotation_0$ be the rotation of 180 degrees around the origin. Let L be the line passing through $(7,0)$ parallel to the y-axis. Find $Rotation_0(L)$. Use your transparency if needed.

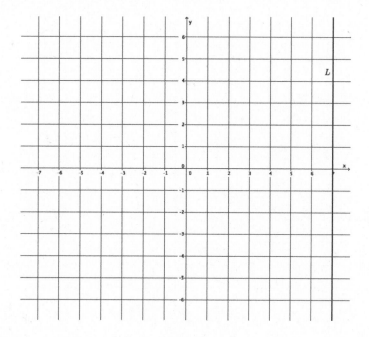

5. Let $Rotation_0$ be the rotation of 180 degrees around the origin. Let L be the line passing through $(0,2)$ parallel to the x-axis. Is L parallel to $Rotation_0(L)$?

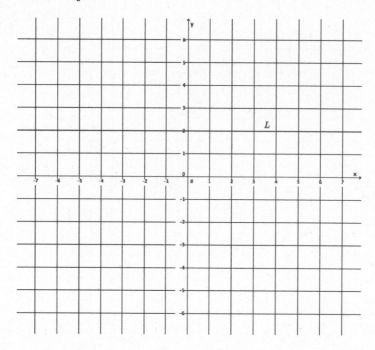

6. Let $Rotation_0$ be the rotation of 180 degrees around the origin. Let line L be the line passing through $(-4,0)$ parallel to the y-axis. Is L parallel to $Rotation_0(L)$?

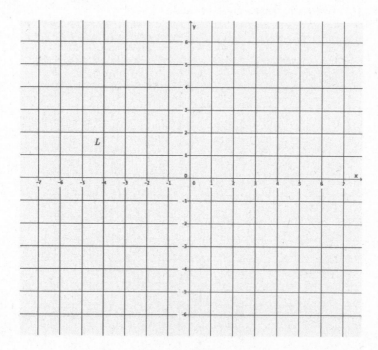

Lesson 6:   Rotations of 180 Degrees

EUREKA
MATH

7. Let $Rotation_0$ be the rotation of 180 degrees around the origin. Let L be the line passing through $(0, -1)$ parallel to the x-axis. Is L parallel to $Rotation_0(L)$?

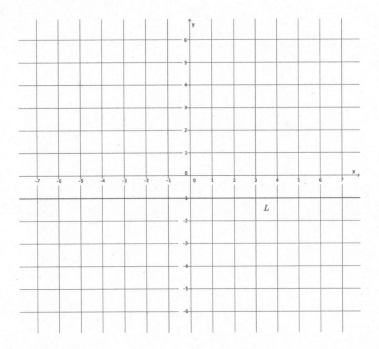

8. Let $Rotation_0$ be the rotation of 180 degrees around the origin. Is L parallel to $Rotation_0(L)$? Use your transparency if needed.

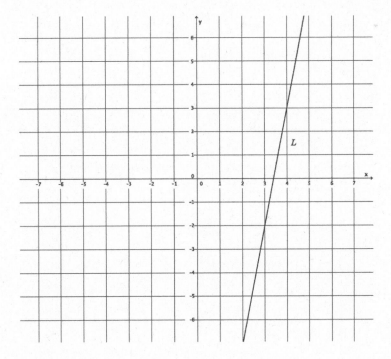

9. Let $Rotation_0$ be the rotation of 180 degrees around the center O. Is L parallel to $Rotation_0(L)$? Use your transparency if needed.

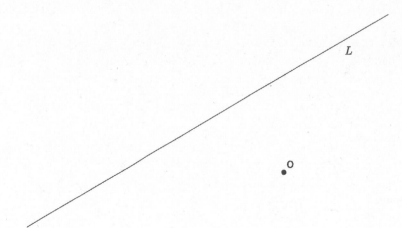

EUREKA MATH®

Lesson Summary

- A rotation of 180 degrees around O is the rigid motion so that if P is any point in the plane, P, O, and *Rotation*(P) are *collinear* (i.e., lie on the same line).

- Given a 180-degree rotation around the origin O of a coordinate system, R_0, and a point P with coordinates (a, b), it is generally said that $R_0(P)$ is the point with coordinates $(-a, -b)$.

THEOREM: Let O be a point not lying on a given line L. Then, the 180-degree rotation around O maps L to a line parallel to L.

Name _____ Date _____

Let there be a rotation of 180 degrees about the origin. Point A has coordinates $(-2, -4)$, and point B has coordinates $(-3, 1)$, as shown below.

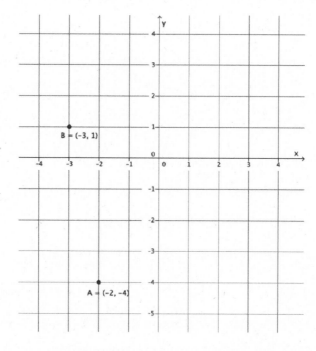

1. What are the coordinates of $Rotation(A)$? Mark that point on the graph so that $Rotation(A) = A'$. What are the coordinates of $Rotation(B)$? Mark that point on the graph so that $Rotation(B) = B'$.

2. What can you say about the points A, A', and O? What can you say about the points B, B', and O?

3. Connect point A to point B to make the line L_{AB}. Connect point A' to point B' to make the line $L_{A'B'}$. What is the relationship between L_{AB} and $L_{A'B'}$?

Lesson Notes

When a line is rotated 180° around a point not on the line, it maps to a line parallel to the given line. A point P with a rotation of 180° around a center O produces a point P' so that P, O, and P' are collinear. When we rotate coordinates 180° around O, the point with coordinates (a, b) is moved to the point with coordinates $(-a, -b)$.

Example

Use the following diagram for Problems 1–5. Use your transparency as needed.

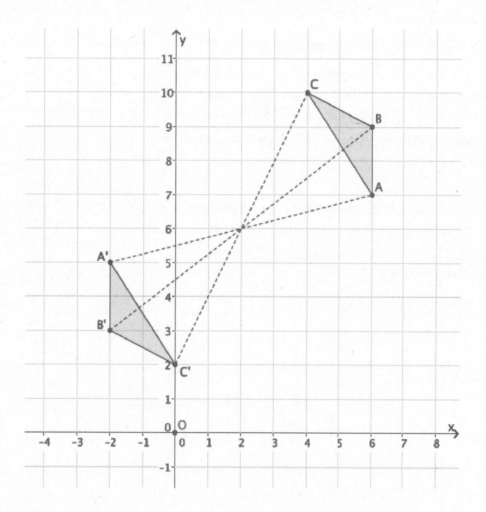

1. Looking only at segment BC, is it possible that a 180° rotation would map segment BC onto segment $B'C'$? Why or why not?

   It is possible because the segments are parallel.

2. Looking only at segment AB, is it possible that a 180° rotation would map segment AB onto segment $A'B'$? Why or why not?

   It is possible because the segments are parallel.

   > I will use my transparency to verify that the segments are parallel. I think the center of rotation is the point $(2, 6)$.

3. Looking only at segment AC, is it possible that a 180° rotation would map segment AC onto segment $A'C'$? Why or why not?

   It is possible because the segments are parallel.

4. Connect point B to point B', point C to point C', and point A to point A'. What do you notice? What do you think that point is?

   All of the lines intersect at one point. The point is the center of rotation. I checked by using my transparency.

   > I checked each segment and its rotated segment to see if they were parallel. I found the center of rotation, so I can say there is a rotation of 180° about a center.

5. Would a rotation map $\triangle ABC$ onto $\triangle A'B'C'$? If so, define the rotation (i.e., degree and center). If not, explain why not.

   Let there be a rotation of 180° around point $(2,6)$. Then, Rotation$(\triangle ABC) = \triangle A'B'C'$.

EUREKA MATH

Use the following diagram for Problems 1–5. Use your transparency as needed.

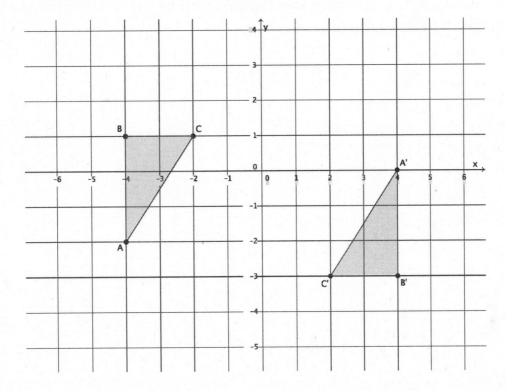

1. Looking only at segment BC, is it possible that a 180° rotation would map segment BC onto segment $B'C'$? Why or why not?

2. Looking only at segment AB, is it possible that a 180° rotation would map segment AB onto segment $A'B'$? Why or why not?

3. Looking only at segment AC, is it possible that a 180° rotation would map segment AC onto segment $A'C'$? Why or why not?

4. Connect point B to point B', point C to point C', and point A to point A'. What do you notice? What do you think that point is?

5. Would a rotation map triangle ABC onto triangle $A'B'C'$? If so, define the rotation (i.e., degree and center). If not, explain why not.

6. The picture below shows right triangles ABC and $A'B'C'$, where the right angles are at B and B'. Given that $AB = A'B' = 1$, and $BC = B'C' = 2$, and that $\overline{AB}$ is not parallel to $\overline{A'B'}$, is there a 180° rotation that would map $\triangle ABC$ onto $\triangle A'B'C'$? Explain.

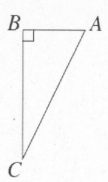
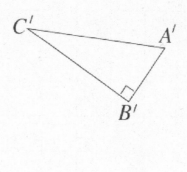

EUREKA
MATH®

Exploratory Challenge

1.

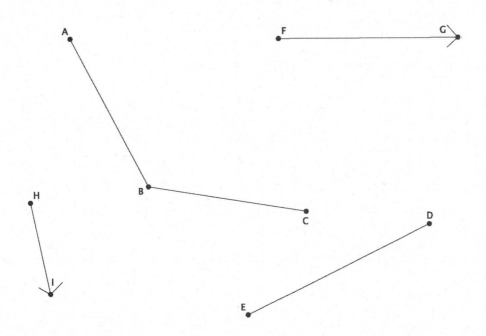

a. Translate ∠*ABC* and segment *ED* along vector $\overrightarrow{FG}$. Label the translated images appropriately, that is, ∠*A'B'C'* and segment *E'D'*.

b. Translate ∠*A'B'C'* and segment *E'D'* along vector $\overrightarrow{HI}$. Label the translated images appropriately, that is, ∠*A"B"C"* and segment *E"D"*.

c. How does the size of ∠*ABC* compare to the size of ∠*A"B"C"*?

d.   How does the length of segment ED compare to the length of the segment $E''D''$?

e.   Why do you think what you observed in parts (d) and (e) were true?

2.   Translate $\triangle\ ABC$ along vector $\overrightarrow{FG}$, and then translate its image along vector $\overrightarrow{JK}$. Be sure to label the images appropriately.

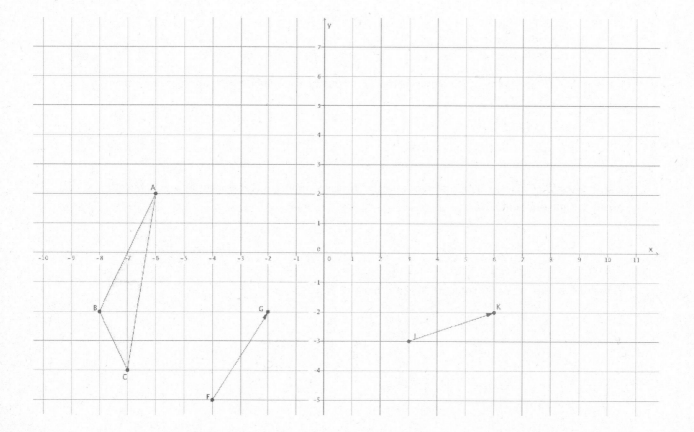

EUREKA
MATH®

3. Translate figure $ABCDEF$ along vector $\overrightarrow{GH}$. Then translate its image along vector $\overrightarrow{JI}$. Label each image appropriately.

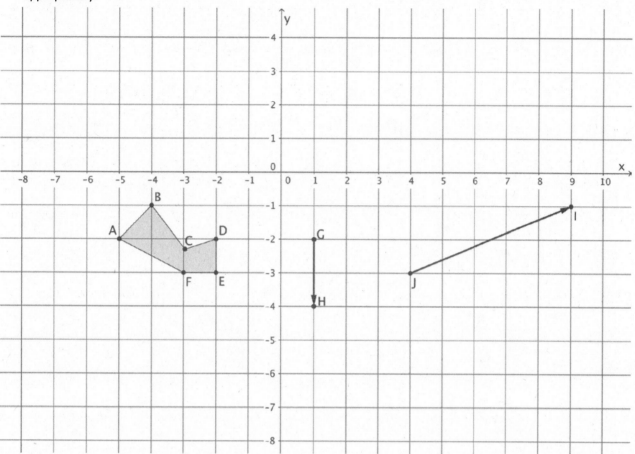

4.

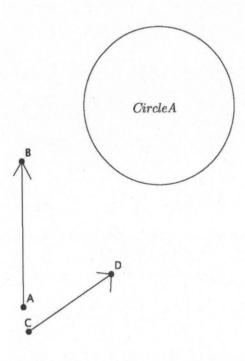

a. Translate Circle A and Ellipse E along vector $\overrightarrow{AB}$. Label the images appropriately.

b. Translate Circle A' and Ellipse E' along vector $\overrightarrow{CD}$. Label each image appropriately.

c. Did the size or shape of either figure change after performing the sequence of translations? Explain.

EUREKA
MATH

5. The picture below shows the translation of Circle A along vector $\overrightarrow{CD}$. Name the vector that maps the image of Circle A back to its original position.

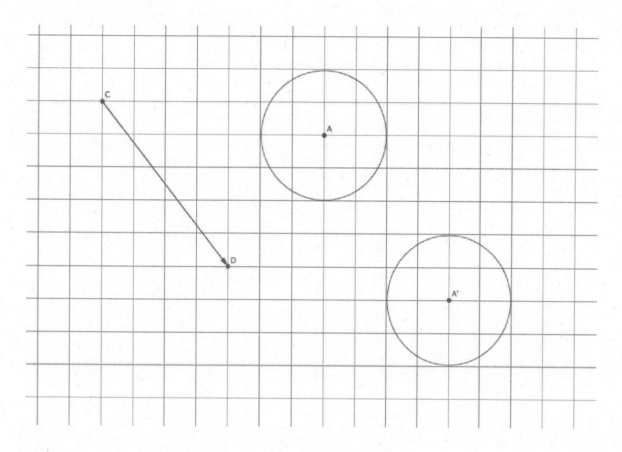

6. If a figure is translated along vector $\overrightarrow{QR}$, what translation takes the figure back to its original location?

Lesson Summary

- Translating a figure along one vector and then translating its image along another vector is an example of a sequence of transformations.

- A sequence of translations enjoys the same properties as a single translation. Specifically, the figures' lengths and degrees of angles are preserved.

- If a figure undergoes two transformations, F and G, and is in the same place it was originally, then the figure has been mapped onto itself.

Name _____   Date _____

Use the picture below to answer Problems 1 and 2.

1. Describe a sequence of translations that would map Figure H onto Figure K.

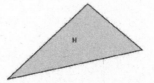
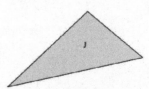
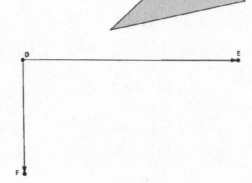

2. Describe a sequence of translations that would map Figure J onto itself.

Lesson Notes

Sequences of translations have the same properties of a single translation (i.e., map lines to lines, rays to rays, segments to segments, and angles to angles). Sequences of translations preserve lengths of segments and measures of angles, in degrees. If a figure undergoes two transformations and is in the same place as it was originally, then the figure has been mapped onto itself.

Examples

1. Sequence translations of rectangle $ABCD$ (a quadrilateral in which both pairs of opposite sides are parallel) along vectors $\overrightarrow{EF}$ and $\overrightarrow{GH}$. Label the translated images.

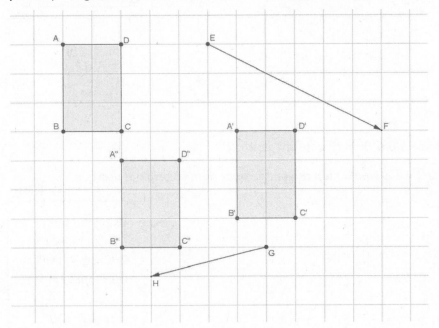

I will trace the rectangle and the vector $\overrightarrow{EF}$ onto my transparency first. Then I will note the image as $A'B'C'D'$. After I have translated the rectangle along vector $\overrightarrow{EF}$, I will trace vector $\overrightarrow{GH}$ and translate the rectangle $A'B'C'D'$, resulting in the final image $A''B''C''D''$.

2. What do you know about $\overline{AD}$ and $\overline{BC}$ compared with $\overline{A'D'}$ and $\overline{B'C'}$? Explain.

   By the definition of a rectangle, $\overline{AD} \parallel \overline{BC}$. Since translations map parallel lines to parallel lines, I know that $\overline{A'D'} \parallel \overline{B'C'}$.

   I remember this from Lesson 3.

3. Are the segments $A'B'$ and $A''B''$ equal in length? How do you know?

   Yes, $|A'B'| = |A''B''|$. Translations preserve lengths of segments.

4. Translate the shape $ABCD$ along the given vector. Label the image.

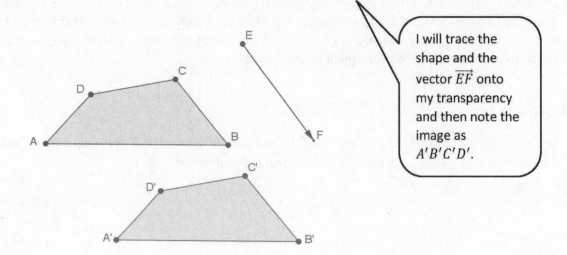

I will trace the shape and the vector $\overrightarrow{EF}$ onto my transparency and then note the image as $A'B'C'D'$.

5. What vector would map the shape $A'B'C'D'$ back onto shape $ABCD$?

Translating the image along vector $\overrightarrow{FE}$ would map the image back onto its original position.

Using the same transparency for Problem 4, I will translate along the vector $\overrightarrow{FE}$ to map $A'B'C'D'$ back to the shape $ABCD$.

EUREKA MATH

1. Sequence translations of parallelogram $ABCD$ (a quadrilateral in which both pairs of opposite sides are parallel) along vectors $\overrightarrow{HG}$ and $\overrightarrow{FE}$. Label the translated images.

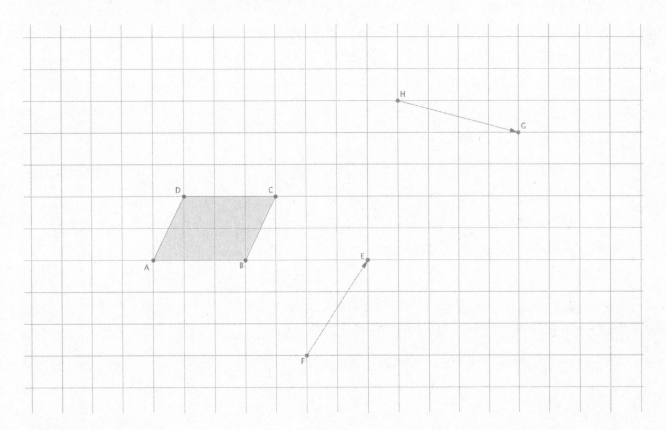

2. What do you know about $\overline{AD}$ and $\overline{BC}$ compared with $\overline{A'D'}$ and $\overline{B'C'}$? Explain.

3. Are the segments $A'B'$ and $A''B''$ equal in length? How do you know?

4. Translate the curved shape ABC along the given vector. Label the image.

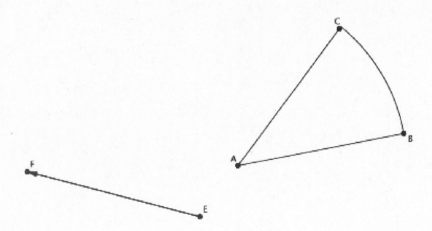

5. What vector would map the shape $A'B'C'$ back onto shape ABC?

EUREKA MATH

Use the figure below to answer Exercises 1–3.

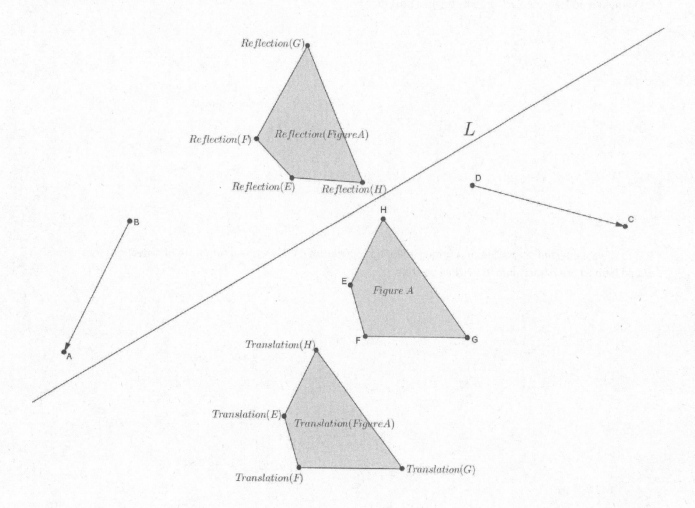

EUREKA
MATH

© 2019 Great Minds®. eureka-math.org

1. Figure A was translated along vector $\overrightarrow{BA}$, resulting in *Translation*(*Figure A*). Describe a sequence of translations that would map Figure A back onto its original position.

2. Figure A was reflected across line L, resulting in *Reflection*(*Figure A*). Describe a sequence of reflections that would map Figure A back onto its original position.

3. Can *Translation*$_{\overrightarrow{BA}}$ of Figure A undo the transformation of *Translation*$_{\overrightarrow{DC}}$ of Figure A? Why or why not?

Lesson 8:   Sequencing Reflections and Translations

EUREKA
MATH

Exercises 4–7

Let S be the black figure.

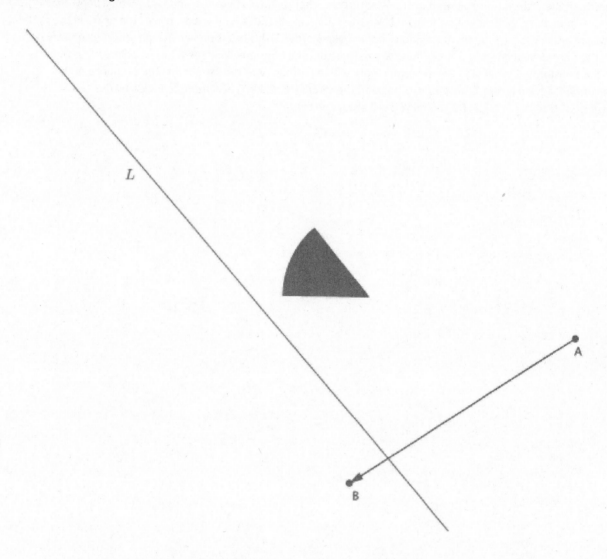

4. Let there be the translation along vector $\overrightarrow{AB}$ and a reflection across line L.

    Use a transparency to perform the following sequence: Translate figure S; then, reflect figure S. Label the image S'.

5. Let there be the translation along vector $\overrightarrow{AB}$ and a reflection across line L.

    Use a transparency to perform the following sequence: Reflect figure S; then, translate figure S. Label the image S''.

6. Using your transparency, show that under a sequence of any two translations, *Translation* and *Translation*$_0$ (along different vectors), that the sequence of the *Translation* followed by the *Translation*$_0$ is equal to the sequence of the *Translation*$_0$ followed by the *Translation*. That is, draw a figure, A, and two vectors. Show that the translation along the first vector, followed by a translation along the second vector, places the figure in the same location as when you perform the translations in the reverse order. (This fact is proven in high school Geometry.) Label the transformed image A'. Now, draw two new vectors and translate along them just as before. This time, label the transformed image A''. Compare your work with a partner. Was the statement "the sequence of the *Translation* followed by the *Translation*$_0$ is equal to the sequence of the *Translation*$_0$ followed by the *Translation*" true in all cases? Do you think it will always be true?

7. Does the same relationship you noticed in Exercise 6 hold true when you replace one of the translations with a reflection. That is, is the following statement true: A translation followed by a reflection is equal to a reflection followed by a translation?

EUREKA
MATH®

Lesson Summary

- A reflection across a line followed by a reflection across the same line places all figures in the plane back onto their original position.

- A reflection followed by a translation does not necessarily place a figure in the same location in the plane as a translation followed by a reflection. The order in which we perform a sequence of rigid motions matters.

Name _____ Date _____

Draw a figure, A, a line of reflection, L, and a vector $\overrightarrow{FG}$ in the space below. Show that under a sequence of a translation and a reflection, that the sequence of the reflection followed by the translation is not equal to the translation followed by the reflection. Label the figure as A' after finding the location according to the sequence reflection followed by the translation, and label the figure A'' after finding the location according to the composition translation followed by the reflection. If A' is not equal to A'', then we have shown that the sequence of the reflection followed by a translation is not equal to the sequence of the translation followed by the reflection. (This is proven in high school Geometry.)

1. Let there be a reflection across line L, and let there be a translation along vector $\overrightarrow{HJ}$. Compare the translated Figure S followed by the reflected image of Figure S with the reflected Figure S followed by the translated image of Figure S. What do you notice?

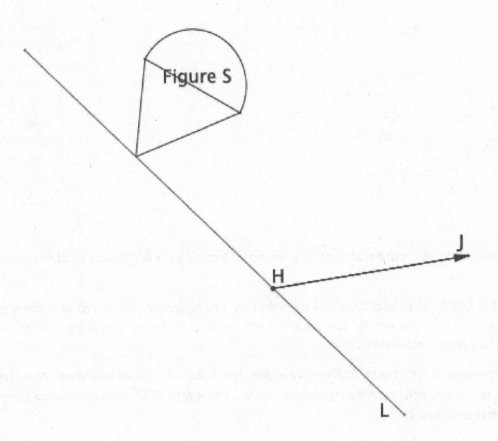

EUREKA
MATH

© 2019 Great Minds®. eureka-math.org

Sample student response.

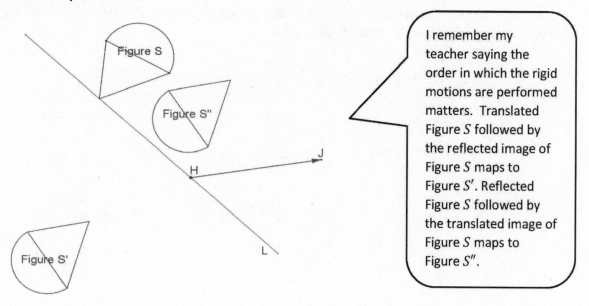

> I remember my teacher saying the order in which the rigid motions are performed matters. Translated Figure S followed by the reflected image of Figure S maps to Figure S'. Reflected Figure S followed by the translated image of Figure S maps to Figure S''.

Students should notice that the two sequences place Figure S in different locations in the plane.

2. Let L_1 and L_2 be parallel lines, and let $Reflection_1$ and $Reflection_2$ be the reflections across L_1 and L_2, respectively (in that order). Can you guess what $Reflection_1$ followed by $Reflection_2$ is? Give as persuasive an argument as you can.

The sequence Reflection$_1$ followed by Reflection$_2$ is just like the translation along a vector $\overrightarrow{EF}$, as shown below, where $\overrightarrow{EF}$ is perpendicular to L_1. The length of $\overrightarrow{EF}$ is equal to twice the distance between L_1 and L_2.

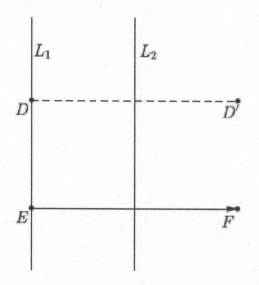

> I'm going to draw a diagram to help me explain. I will pick a point D on L_1 and reflect it across line L_1 first, and then I'll reflect across line L_2. When I did the first reflection, point D stayed on L_1 since it is on the line of reflection. When I did the second reflection, I noticed that the point D' looked like it had been translated along a vector. I checked with my transparency.

1. Let there be a reflection across line L, and let there be a translation along vector $\overrightarrow{AB}$, as shown. If S denotes the black figure, compare the translated figure S followed by the reflected image of figure S with the reflected figure S followed by the translated image of figure S.

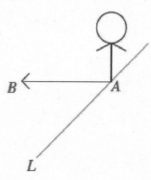

2. Let L_1 and L_2 be parallel lines, and let $Reflection_1$ and $Reflection_2$ be the reflections across L_1 and L_2, respectively (in that order). Show that a $Reflection_2$ followed by $Reflection_1$ is not equal to a $Reflection_1$ followed by $Reflection_2$. (Hint: Take a point on L_1 and see what each of the sequences does to it.)

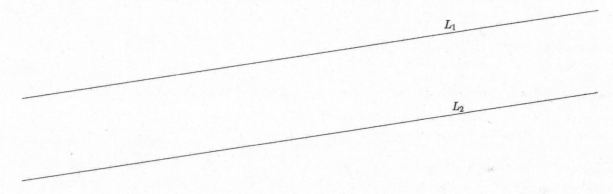

3. Let L_1 and L_2 be parallel lines, and let $Reflection_1$ and $Reflection_2$ be the reflections across L_1 and L_2, respectively (in that order). Can you guess what $Reflection_1$ followed by $Reflection_2$ is? Give as persuasive an argument as you can. (Hint: Examine the work you just finished for the last problem.)

Exploratory Challenge

1.

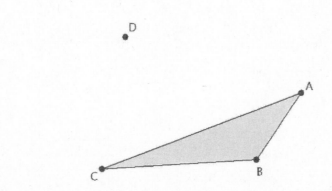

a. Rotate △ ABC d degrees around center D. Label the rotated image as △ $A'B'C'$.

b. Rotate △ $A'B'C'$ d degrees around center E. Label the rotated image as △ $A''B''C''$.

c. Measure and label the angles and side lengths of △ ABC. How do they compare with the images △ $A'B'C'$ and △ $A''B''C''$?

d. How can you explain what you observed in part (c)? What statement can you make about properties of sequences of rotations as they relate to a single rotation?

EUREKA MATH®

2.

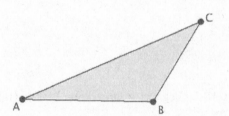

a. Rotate △ ABC d degrees around center D, and then rotate again d degrees around center E. Label the image as △ $A'B'C'$ after you have completed both rotations.

b. Can a single rotation around center D map △ $A'B'C'$ onto △ ABC?

c. Can a single rotation around center E map △ $A'B'C'$ onto △ ABC?

d. Can you find a center that would map △ $A'B'C'$ onto △ ABC in one rotation? If so, label the center F.

EUREKA
MATH®

1. Let there be a reflection across line L, and let there be a translation along vector $\overrightarrow{AB}$, as shown. If S denotes the black figure, compare the translated figure S followed by the reflected image of figure S with the reflected figure S followed by the translated image of figure S.

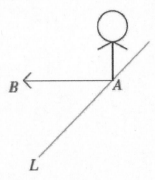

2. Let L_1 and L_2 be parallel lines, and let $Reflection_1$ and $Reflection_2$ be the reflections across L_1 and L_2, respectively (in that order). Show that a $Reflection_2$ followed by $Reflection_1$ is not equal to a $Reflection_1$ followed by $Reflection_2$. (Hint: Take a point on L_1 and see what each of the sequences does to it.)

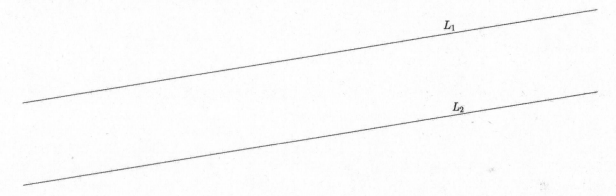

3. Let L_1 and L_2 be parallel lines, and let $Reflection_1$ and $Reflection_2$ be the reflections across L_1 and L_2, respectively (in that order). Can you guess what $Reflection_1$ followed by $Reflection_2$ is? Give as persuasive an argument as you can. (Hint: Examine the work you just finished for the last problem.)

3.

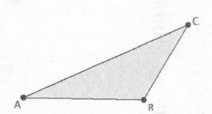

a.   Rotate △ ABC 90° (counterclockwise) around center D, and then rotate the image another 90° (counterclockwise) around center E. Label the image △ $A'B'C'$.

b.   Rotate △ ABC 90° (counterclockwise) around center E, and then rotate the image another 90° (counterclockwise) around center D. Label the image △ $A''B''C''$.

c.   What do you notice about the locations of △ $A'B'C'$ and △ $A''B''C''$? Does the order in which you rotate a figure around different centers have an impact on the final location of the figure's image?

EUREKA
MATH®

4.

a. Rotate △ ABC 90° (counterclockwise) around center D, and then rotate the image another 45° (counterclockwise) around center D. Label the image △ $A'B'C'$.

b. Rotate △ ABC 45° (counterclockwise) around center D, and then rotate the image another 90° (counterclockwise) around center D. Label the image △ $A''B''C''$.

c. What do you notice about the locations of △ $A'B'C'$ and △ $A''B''C''$? Does the order in which you rotate a figure around the same center have an impact on the final location of the figure's image?

5. △ *ABC* has been rotated around two different centers, and its image is △ *A'B'C'*. Describe a sequence of rigid motions that would map △ *ABC* onto △ *A'B'C'*.

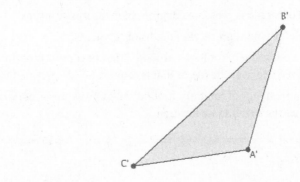

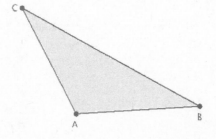

EUREKA
MATH®

Lesson Summary

- Sequences of rotations have the same properties as a single rotation:

  - A sequence of rotations preserves degrees of measures of angles.

  - A sequence of rotations preserves lengths of segments.

- The order in which a sequence of rotations around different centers is performed matters with respect to the final location of the image of the figure that is rotated.

- The order in which a sequence of rotations around the same center is performed does not matter. The image of the figure will be in the same location.

EUREKA MATH

Name _____   Date _____

1. Let *Rotation*₁ be the rotation of a figure *d* degrees around center *O*. Let *Rotation*₂ be the rotation of the same figure *d* degrees around center *P*. Does the *Rotation*₁ of the figure followed by the *Rotation*₂ equal a *Rotation*₂ of the figure followed by the *Rotation*₁? Draw a picture if necessary.

2. Angle *ABC* underwent a sequence of rotations. The original size of ∠*ABC* is 37°. What was the size of the angle after the sequence of rotations? Explain.

3. Triangle *ABC* underwent a sequence of rotations around two different centers. Its image is △ *A'B'C'*. Describe a sequence of rigid motions that would map △ *ABC* onto △ *A'B'C'*.

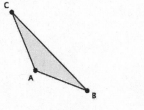
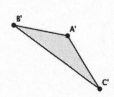

Refer to the figure below to answer Problems 1–3. Note: Figure is not drawn to scale.

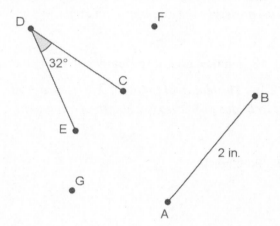

1. Rotate $\angle CDE$ and segment AB d degrees around center F and then d_1 degrees around center G. Label the final location of the images as $\angle C'D'E'$ and segment $A'B'$.

    Drawings will vary based on students' choice of degree to rotate. Shown below is a rotation around point F of 45° followed by a rotation around point G of 80°.

> The order in which I rotate matters because I am using two different centers.

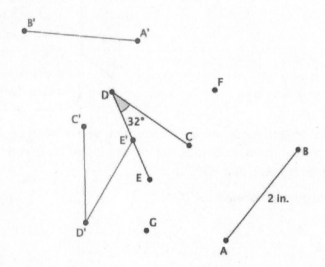

2. What is the measure of ∠CDE, and how does it compare to the measure of ∠C'D'E'? Explain.

    The measure of ∠CDE is 32°. The measure of ∠C'D'E' is 32°. The angles are equal in measure because a sequence of rotations preserves the degrees of an angle.

3. What is the length of segment AB, and how does it compare to the length of segment A'B'? Explain.

    The length of segment AB is 2 in. The length of segment A'B' is also 2 in. The segments are equal in length because a sequence of rotations preserves the length of a segment.

Refer to the figure below to answer Problem 4.

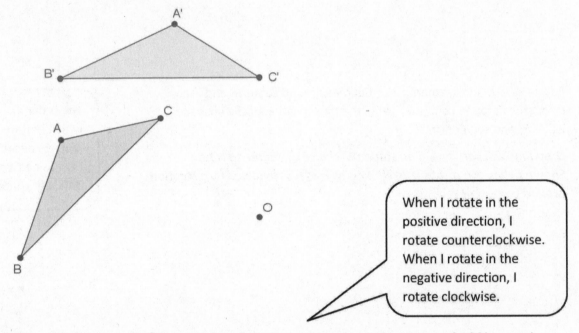

When I rotate in the positive direction, I rotate counterclockwise. When I rotate in the negative direction, I rotate clockwise.

4. Let *Rotation₁* be a rotation of 45° around the center O. Let *Rotation₂* be a rotation of −90° around the same center O. Determine the approximate location of *Rotation₁*(△ ABC) followed by *Rotation₂*(△ ABC). Label the image of △ ABC as △ A'B'C'.

    The image of △ ABC is shown above and labeled △ A'B'C'.

Lesson 9:   Sequencing Rotations   EUREKA MATH

1. Refer to the figure below.

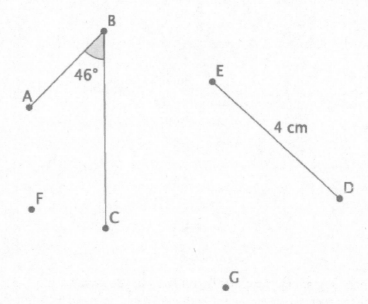

a. Rotate ∠ABC and segment DE d degrees around center F and then d degrees around center G. Label the final location of the images as ∠A'B'C' and segment D'E'.

b. What is the size of ∠ABC, and how does it compare to the size of ∠A'B'C'? Explain.

c. What is the length of segment DE, and how does it compare to the length of segment D'E'? Explain.

2. Refer to the figure given below.

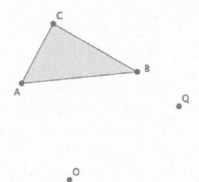

a. Let *Rotation*₁ be a counterclockwise rotation of 90° around the center O. Let *Rotation*₂ be a clockwise rotation of (−45)° around the center Q. Determine the approximate location of *Rotation*₁ (△ ABC) followed by *Rotation*₂. Label the image of △ ABC as △ A'B'C'.

b. Describe the sequence of rigid motions that would map △ ABC onto △ A'B'C'.

3. Refer to the figure given below.

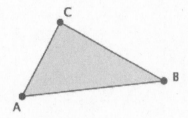

Let R be a rotation of $(-90)°$ around the center O. Let $Rotation_2$ be a rotation of $(-45)°$ around the same center O. Determine the approximate location of $Rotation_1 (\triangle ABC)$ followed by $Rotation_2 (\triangle ABC)$. Label the image of $\triangle ABC$ as $\triangle A'B'C'$.

EUREKA
MATH®

1. In the following picture, triangle ABC can be traced onto a transparency and mapped onto triangle $A'B'C'$.
    Which basic rigid motion, or sequence of, would map one triangle onto the other?

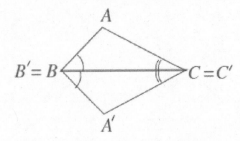

2. In the following picture, triangle ABC can be traced onto a transparency and mapped onto triangle $A'B'C'$.
    Which basic rigid motion, or sequence of, would map one triangle onto the other?

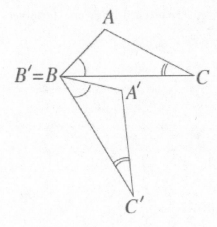

3.   In the following picture, triangle ABC can be traced onto a transparency and mapped onto triangle $A'B'C'$.

Which basic rigid motion, or sequence of, would map one triangle onto the other?

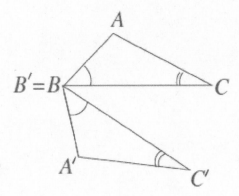

4.   In the following picture, we have two pairs of triangles. In each pair, triangle ABC can be traced onto a transparency and mapped onto triangle $A'B'C'$.

Which basic rigid motion, or sequence of, would map one triangle onto the other?

Scenairo 1:

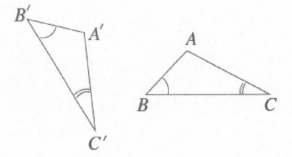

Scenairo 2:

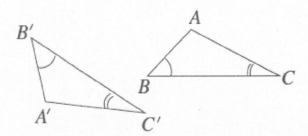

EUREKA
MATH

5. Let two figures ABC and $A'B'C'$ be given so that the length of curved segment AC equals the length of curved segment $A'C'$, $|\angle B| = |\angle B'| = 80°$, and $|AB| = |A'B'| = 5$. With clarity and precision, describe a sequence of rigid motions that would map figure ABC onto figure $A'B'C'$.

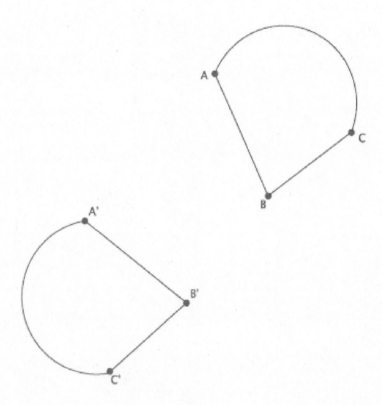

Name _____ Date _____

Triangle ABC has been moved according to the following sequence: a translation followed by a rotation followed by a reflection. With precision, describe each rigid motion that would map $\triangle\ ABC$ onto $\triangle\ A'B'C'$. Use your transparency and add to the diagram if needed.

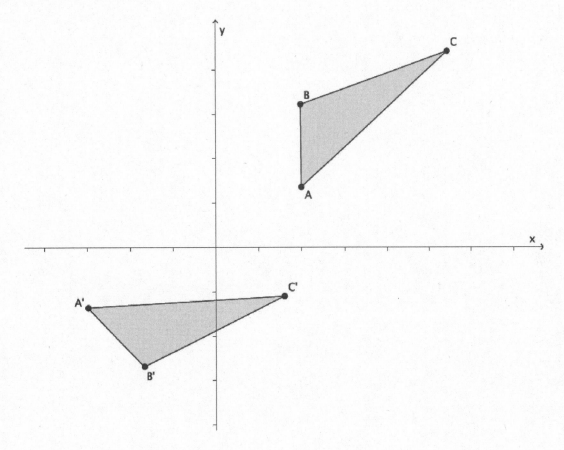

1. Let there be a reflection across the y-axis, let there be a translation along vector $\vec{u}$, and let there be a rotation around point A, 90° (counterclockwise). Let S be the figure as shown below. Show the location of S after performing the following sequence: a reflection followed by a translation followed by a rotation. Label the image as Figure S'.

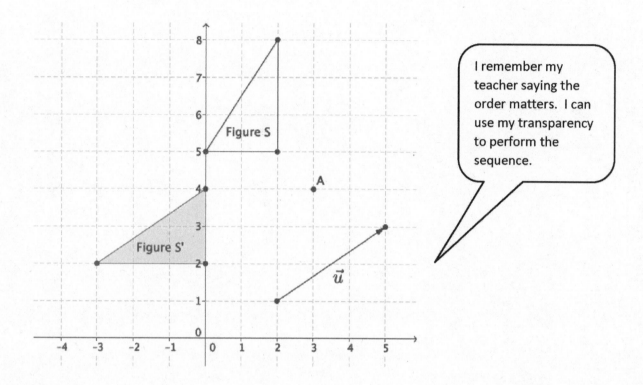

Figure S

A

Figure S'

$\vec{u}$

I remember my teacher saying the order matters. I can use my transparency to perform the sequence.

2. Would the location of the image of S in the previous problem be the same if the translation was performed last instead of second; that is, does the sequence, reflection followed by a rotation followed by a translation, equal a reflection followed by a translation followed by a rotation? Explain.

    No, the order of the transformations matters. If the translation was performed last, the location of the image of S, after the sequence, would be in a different location than if the translation was performed second.

1. Let there be the translation along vector $\vec{v}$, let there be the rotation around point A, −90 degrees (clockwise), and let there be the reflection across line L. Let S be the figure as shown below. Show the location of S after performing the following sequence: a translation followed by a rotation followed by a reflection.

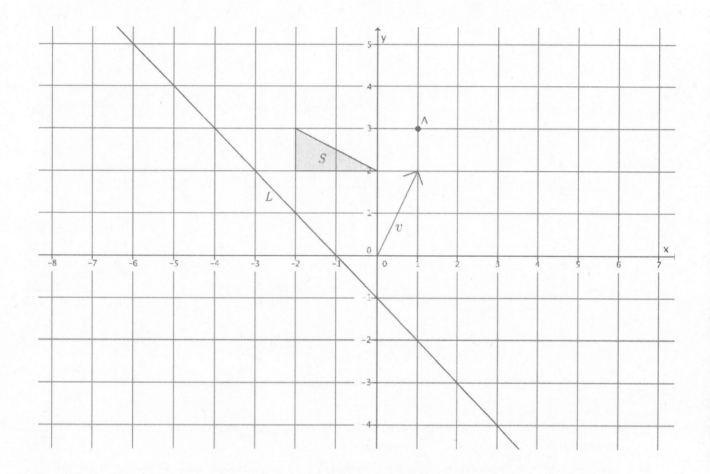

2. Would the location of the image of S in the previous problem be the same if the translation was performed last instead of first; that is, does the sequence, translation followed by a rotation followed by a reflection, equal a rotation followed by a reflection followed by a translation? Explain.

3. Use the same coordinate grid to complete parts (a)–(c).

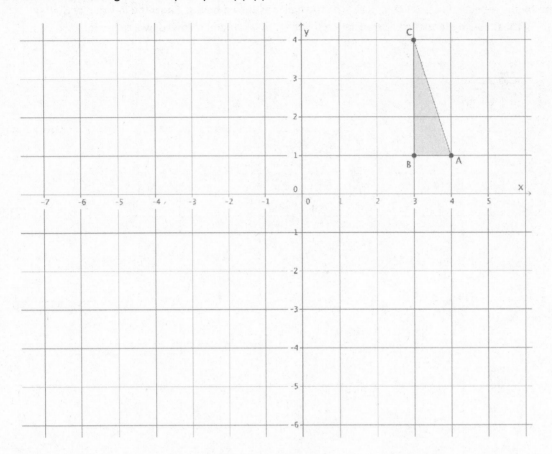

a. Reflect triangle ABC across the vertical line, parallel to the y-axis, going through point $(1, 0)$. Label the transformed points A, B, C as A', B', C', respectively.

b. Reflect triangle $A'B'C'$ across the horizontal line, parallel to the x-axis going through point $(0, -1)$. Label the transformed points of A', B', C' as A'', B'', C'', respectively.

c. Is there a single rigid motion that would map triangle ABC to triangle $A''B''C''$?

EUREKA
MATH®

Exercise 1

a. Describe the sequence of basic rigid motions that shows $S_1 \cong S_2$.

EUREKA
MATH®

© 2019 Great Minds®. eureka-math.org

b. Describe the sequence of basic rigid motions that shows $S_2 \cong S_3$.

EUREKA
MATH

c. Describe a sequence of basic rigid motions that shows $S_1 \cong S_3$.

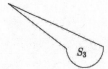

Exercise 2

Perform the sequence of a translation followed by a rotation of Figure XYZ, where T is a translation along a vector $\overrightarrow{AB}$, and R is a rotation of d degrees (you choose d) around a center O. Label the transformed figure $X'\ Y'\ Z'$.

Is $XYZ \cong X'\ Y'\ Z'$?

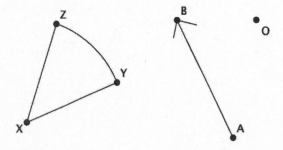

EUREKA
MATH

Lesson Summary

Given that sequences enjoy the same basic properties of basic rigid motions, we can state three basic properties of congruences:

    (Congruence 1)   A congruence maps a line to a line, a ray to a ray, a segment to a segment, and an angle to an angle.

    (Congruence 2)   A congruence preserves lengths of segments.

    (Congruence 3)   A congruence preserves measures of angles.

The notation used for congruence is ≅.

Name _____   Date _____

1. Is $\triangle ABC \cong \triangle A'B'C'$? If so, describe a sequence of rigid motions that proves they are congruent. If not, explain how you know.

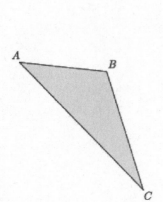
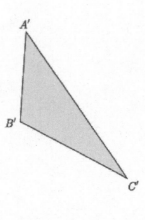

2. Is $\triangle ABC \cong \triangle A'B'C'$? If so, describe a sequence of rigid motions that proves they are congruent. If not, explain how you know.

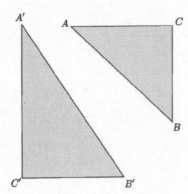

Are the two parallelograms shown below congruent? If so, describe a congruence that would map one parallelogram onto the other.

> To prove two figures are congruent, I have to show that one figure will map onto another using a sequence of rigid motions. I could try it first with my transparency.

> I remember my teacher saying it makes more sense to translate the figure along a vector first to get a common point and then rotate the figure about the point to get a common side that I can then use as the line of reflection.

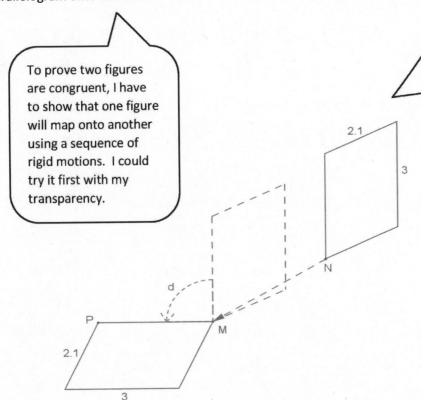

> The diagram doesn't have any points noted. If I'm going to be precise, I'll have to add the points N, M, and P to the drawing.

Sample student response: *Yes, they are congruent. Let there be a translation along vector $\overrightarrow{NM}$. Let there be a rotation around point M, d degrees. Let there be a reflection across line MP. Then, the translation followed by the rotation followed by the reflection will map the parallelogram on the right to the parallelogram on the left.*

1. Given two right triangles with lengths shown below, is there one basic rigid motion that maps one to the other? Explain.

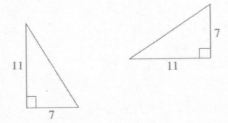

2. Are the two right triangles shown below congruent? If so, describe a congruence that would map one triangle onto the other.

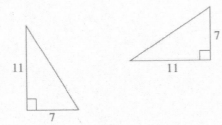

3. Given two rays, $\overrightarrow{OA}$ and $\overrightarrow{O'A'}$:

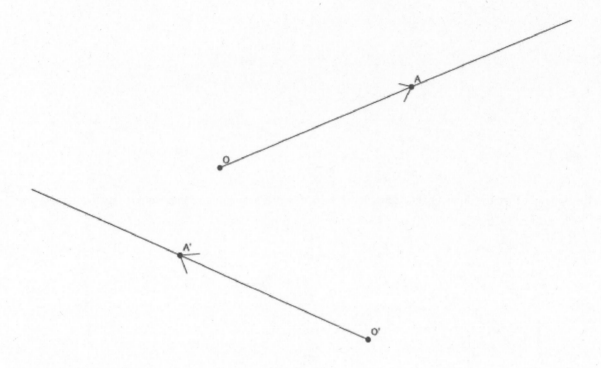

a. Describe a congruence that maps $\overrightarrow{OA}$ to $\overrightarrow{O'A'}$.

b. Describe a congruence that maps $\overrightarrow{O'A'}$ to $\overrightarrow{OA}$.

Lesson 11:   Definition of Congruence and Some Basic Properties

EUREKA
MATH

Exploratory Challenge 1

In the figure below, L_1 is not parallel to L_2, and m is a transversal. Use a protractor to measure angles 1–8. Which, if any, are equal in measure? Explain why. (Use your transparency if needed.)

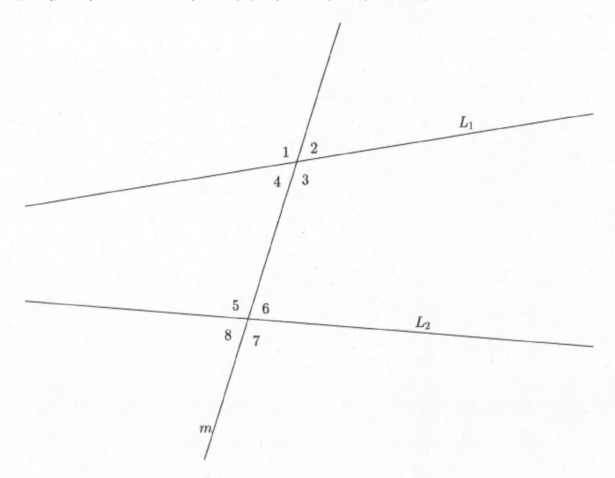

Exploratory Challenge 2

In the figure below, $L_1 \parallel L_2$, and m is a transversal. Use a protractor to measure angles 1–8. List the angles that are equal in measure.

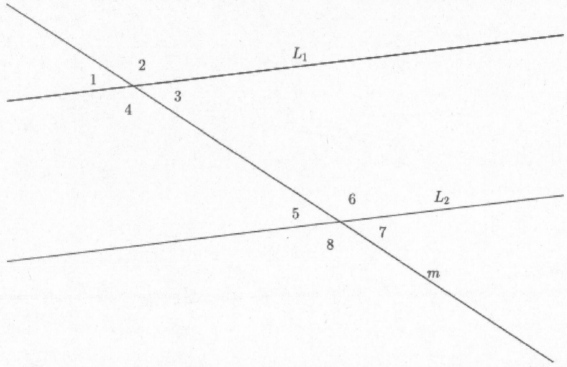

a. What did you notice about the measures of $\angle 1$ and $\angle 5$? Why do you think this is so? (Use your transparency if needed.)

b. What did you notice about the measures of $\angle 3$ and $\angle 7$? Why do you think this is so? (Use your transparency if needed.) Are there any other pairs of angles with this same relationship? If so, list them.

c. What did you notice about the measures of $\angle 4$ and $\angle 6$? Why do you think this is so? (Use your transparency if needed.) Is there another pair of angles with this same relationship?

Lesson Summary

Angles that are on the same side of the transversal in corresponding positions (above each of L_1 and L_2 or below each of L_1 and L_2) are called *corresponding angles*. For example, $\angle 2$ and $\angle 4$ are corresponding angles.

When angles are on opposite sides of the transversal and between (inside) the lines L_1 and L_2, they are called *alternate interior angles*. For example, $\angle 3$ and $\angle 7$ are alternate interior angles.

When angles are on opposite sides of the transversal and outside of the lines (above L_1 and below L_2), they are called *alternate exterior angles*. For example, $\angle 1$ and $\angle 5$ are alternate exterior angles.

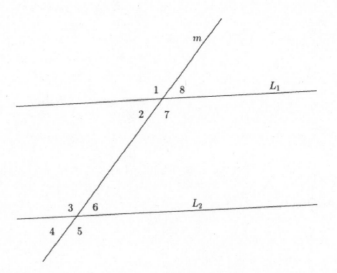

When parallel lines are cut by a transversal, any corresponding angles, any alternate interior angles, and any alternate exterior angles are equal in measure. If the lines are not parallel, then the angles are not equal in measure.

Name _____   Date _____

Use the diagram to answer Questions 1 and 2. In the diagram, lines L_1 and L_2 are intersected by transversal m, forming angles 1–8, as shown.

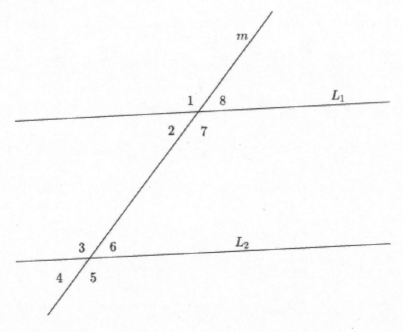

1. If $L_1 \parallel L_2$, what do you know about $\angle 2$ and $\angle 6$? Use informal arguments to support your claim.

2. If $L_1 \parallel L_2$, what do you know about $\angle 1$ and $\angle 3$? Use informal arguments to support your claim.

Use the diagram below to complete Problems 1–2.

Even though lines L_1 and L_2 might look parallel, I can't assume that they are. Since line m crosses lines L_1 and L_2, then line m is the transversal.

1. Identify all pairs of corresponding angles. Are the pairs of corresponding angles equal in measure? How do you know?

   ∠1 and ∠3, ∠2 and ∠4, ∠8 and ∠6, ∠7 and ∠5,

   There is no information provided about the lines in the diagram being parallel. For that reason, we do not know if the pairs of corresponding angles are equal in measure. If we knew the lines were parallel, we could use a translation to map one angle onto another.

   Corresponding angles are on the same side of the transversal in corresponding positions.

2. Identify all pairs of alternate interior angles. Are the pairs of alternate interior angles equal in measure? How do you know?

   ∠2 and ∠6, ∠3 and ∠7,

   There is no information provided about the lines in the diagram being parallel. For that reason, we do not know if the pairs of alternate interior angles are equal in measure. If the lines were parallel, we could use a rotation to show that the pairs of angles would map onto one another, proving they are equal in measure.

   Alternate interior angles are on opposite sides of the transversal in between lines L_1 and L_2.

Use the diagram below to complete Problems 3–4, in the diagram, $L_1 \parallel L_2$.

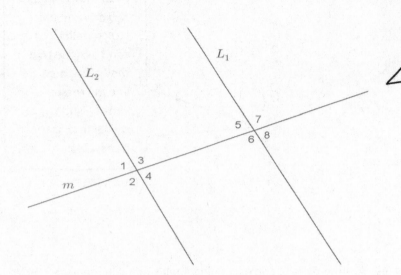

> When parallel lines are cut by a transversal, pairs of corresponding angles, alternate interior angles, and alternate exterior angles are equal in measure.

3. Use an informal argument to describe why ∠1 and ∠8 are equal in measure.

   The reason that ∠1 and ∠8 are equal in measure when the lines are parallel is because you can rotate around the midpoint of the segment between the parallel lines. A rotation would then map ∠1 onto ∠8, showing that they are congruent and equal in measure.

   > I remember my teacher marking the midpoint between lines L_1 and L_2 on line m and demonstrating the rotation. I remember from Lesson 5 that rotation preserves degrees of angles.

4. Use an informal argument to describe why ∠1 and ∠5 are equal in measure.

   The reason that ∠1 and ∠5 are equal in measure when the lines are parallel is because you can translate along a vector equal in length of the segment between the parallel lines; then, ∠1 would map onto ∠5.

   > I remember from Lesson 2 that translations along a vector preserve degrees of measures of angles. My teacher demonstrated this in class by translating along a vector on line m the exact distance between lines L_1 and L_2.

EUREKA MATH

Use the diagram below to do problem 1–10.

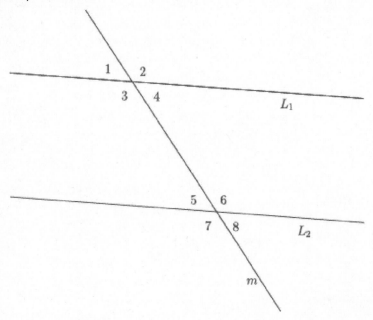

1. Identify all pairs of corresponding angles. Are the pairs of corresponding angles equal in measure? How do you know?

2. Identify all pairs of alternate interior angles. Are the pairs of alternate interior angles equal in measure? How do you know?

3. Use an informal argument to describe why ∠1 and ∠8 are equal in measure if $L_1 \parallel L_2$.

4. Assuming $L_1 \parallel L_2$, if the measure of ∠4 is 73°, what is the measure of ∠8? How do you know?

5. Assuming $L_1 \parallel L_2$, if the measure of ∠3 is 107° degrees, what is the measure of ∠6? How do you know?

6. Assuming $L_1 \parallel L_2$, if the measure of ∠2 is 107°, what is the measure of ∠7? How do you know?

7. Would your answers to Problems 4–6 be the same if you had not been informed that $L_1 \parallel L_2$? Why or why not?

8. Use an informal argument to describe why ∠1 and ∠5 are equal in measure if $L_1 \parallel L_2$.

9. Use an informal argument to describe why ∠4 and ∠5 are equal in measure if $L_1 \parallel L_2$.

10. Assume that L_1 is not parallel to L_2. Explain why ∠3 ≠ ∠7.

Concept Development

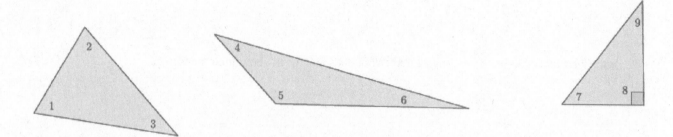

$$m\angle 1 + m\angle 2 + m\angle 3 = m\angle 4 + m\angle 5 + m\angle 6 = m\angle 7 + m\angle 8 + m\angle 9 = 180°$$

Note that the sum of the measures of angles 7 and 9 must equal 90° because of the known right angle in the right triangle.

Let triangle ABC be given. On the ray from B to C, take a point D so that C is between B and D. Through point C, draw a segment parallel to $\overline{AB}$, as shown. Extend the segments AB and CE. Line AC is the transversal that intersects the parallel lines.

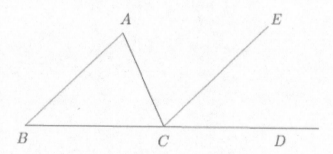

a. Name the three interior angles of triangle ABC.

b. Name the straight angle.

EUREKA
MATH®

c.   What kinds of angles are $\angle ABC$ and $\angle ECD$? What does that mean about their measures?

d.   What kinds of angles are $\angle BAC$ and $\angle ECA$? What does that mean about their measures?

e.   We know that $m\angle BCD = m\angle BCA + m\angle ECA + m\angle ECD = 180°$. Use substitution to show that the measures of the three interior angles of the triangle have a sum of $180°$.

Exploratory Challenge 2

The figure below shows parallel lines L_1 and L_2. Let m and n be transversals that intersect L_1 at points B and C, respectively, and L_2 at point F, as shown. Let A be a point on L_1 to the left of B, D be a point on L_1 to the right of C, G be a point on L_2 to the left of F, and E be a point on L_2 to the right of F.

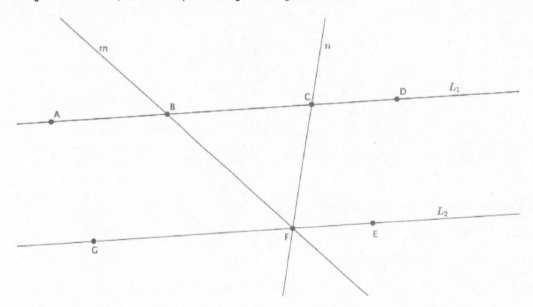

a.   Name the triangle in the figure.

EUREKA
MATH®

b. Name a straight angle that will be useful in proving that the sum of the measures of the interior angles of the triangle is 180°.

c. Write your proof below.

Lesson Summary

All triangles have a sum of measures of the interior angles equal to 180°.

The proof that a triangle has a sum of measures of the interior angles equal to 180° is dependent upon the knowledge of straight angles and angle relationships of parallel lines cut by a transversal.

Lesson 13:   Angle Sum of a Triangle

Name _____   Date _____

1. If $L_1 \parallel L_2$, and $L_3 \parallel L_4$, what is the measure of $\angle 1$? Explain how you arrived at your answer.

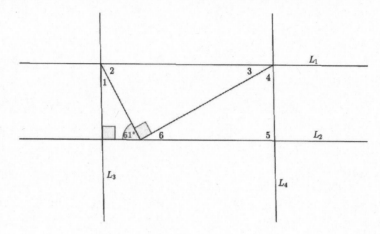

2. Given that line AB is parallel to line CE, present an informal argument to prove that the measures of the interior angles of triangle ABC have a sum of 180°.

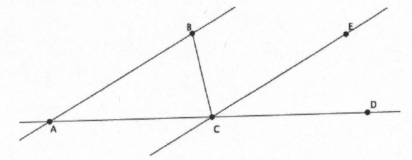

EUREKA MATH

© 2019 Great Minds®. eureka-math.org

1. In the diagram below, line AB is parallel to line EF; that is, $L_{AB} \parallel L_{EF}$. The measure of $\angle BAH$ is 21°, and the measure of $\angle FEH$ is 36°. Find the measure of $\angle AHE$. Explain why you are correct by presenting an informal argument that uses the angle sum of a triangle. (Hint: Extend the segment EH so that it intersects line AB.)

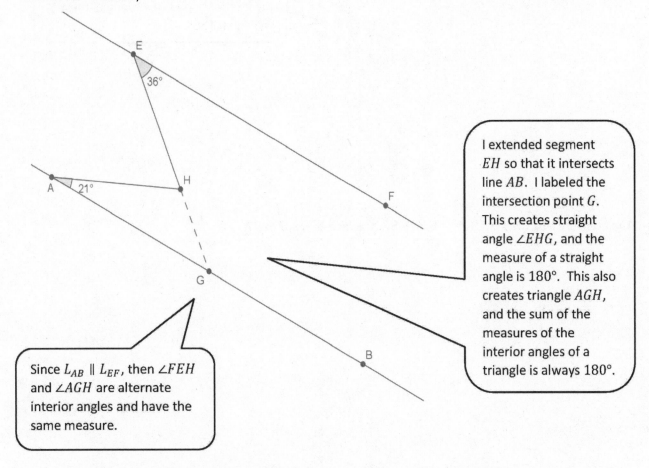

I extended segment EH so that it intersects line AB. I labeled the intersection point G. This creates straight angle $\angle EHG$, and the measure of a straight angle is 180°. This also creates triangle AGH, and the sum of the measures of the interior angles of a triangle is always 180°.

Since $L_{AB} \parallel L_{EF}$, then $\angle FEH$ and $\angle AGH$ are alternate interior angles and have the same measure.

Since $\angle FEH$ and $\angle AGH$ are alternate interior angles of parallel lines, the angles are congruent and have the same measure. Since the angle sum of a triangle is 180°, the measure of $\angle AHG$ is $180° - (36° + 21°)$, which is equal to 123°. The straight angle $\angle EHG$ is made up of $\angle AHG$ and $\angle AHE$. Since straight angles measure 180° and the measure of $\angle AHG$ is 123°, then the measure of $\angle AHE$ is 57°.

2. What is the measure of ∠*CAB*?

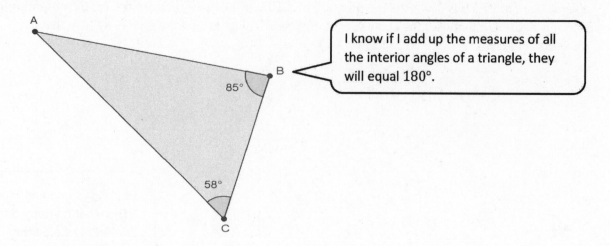

I know if I add up the measures of all the interior angles of a triangle, they will equal 180°.

The measure of ∠CAB is $180° - (85° + 58°)$, *which is equal to* $37°$.

Lesson 13:   Angle Sum of a Triangle

EUREKA
MATH

1. In the diagram below, line AB is parallel to line CD, that is, $L_{AB} \parallel L_{CD}$. The measure of $\angle ABC$ is 28°, and the measure of $\angle EDC$ is 42°. Find the measure of $\angle CED$. Explain why you are correct by presenting an informal argument that uses the angle sum of a triangle.

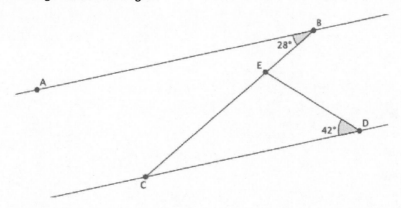

2. In the diagram below, line AB is parallel to line CD, that is, $L_{AB} \parallel L_{CD}$. The measure of $\angle ABE$ is 38°, and the measue of $\angle EDC$ is 16°. Find the measure of $\angle BED$. Explain why you are correct by presenting an informal argument that uses the angle sum of a triangle. (Hint: Find the measure of $\angle CED$ first, and then use that measure to find the measure of $\angle BED$.)

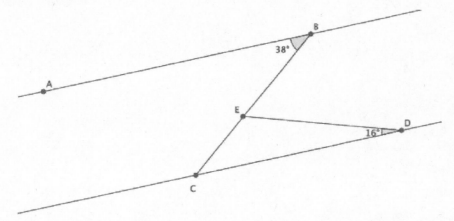

3. In the diagram below, line AB is parallel to line CD, that is, $L_{AB} \parallel L_{CD}$. The measure of $\angle ABE$ is 56°, and the measure of $\angle EDC$ is 22°. Find the measure of $\angle BED$. Explain why you are correct by presenting an informal argument that uses the angle sum of a triangle. (Hint: Extend the segment BE so that it intersects line CD.)

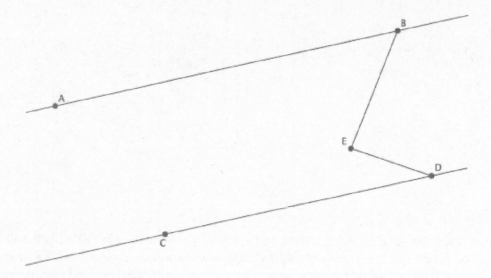

4. What is the measure of $\angle ACB$?

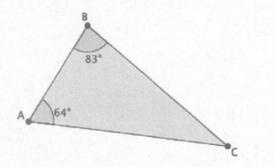

5. What is the measure of $\angle EFD$?

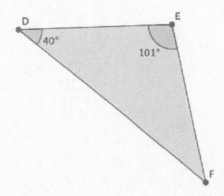

Lesson 13:   Angle Sum of a Triangle

EUREKA
MATH®

6. What is the measure of ∠*HIG*?

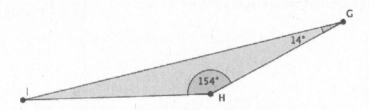

7. What is the measure of ∠*ABC*?

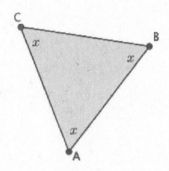

8. Triangle *DEF* is a right triangle. What is the measure of ∠*EFD*?

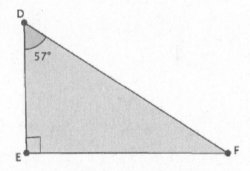

9. In the diagram below, Lines L_1 and L_2 are parallel. Transversals r and s intersect both lines at the points shown below. Determine the measure of $\angle JMK$. Explain how you know you are correct.

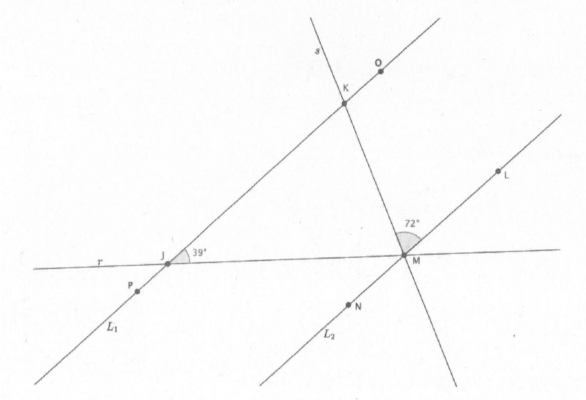

EUREKA MATH

Exercises 1–4

Use the diagram below to complete Exercises 1–4.

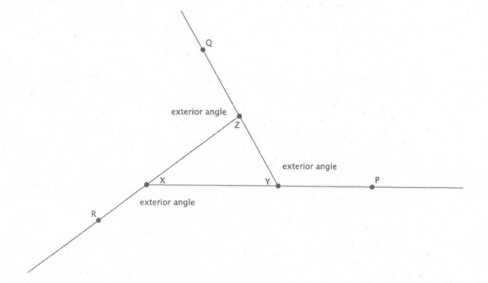

1. Name an exterior angle and the related remote interior angles.

2. Name a second exterior angle and the related remote interior angles.

3. Name a third exterior angle and the related remote interior angles.

4. Show that the measure of an exterior angle is equal to the sum of the measures of the related remote interior angles.

Example 1

Find the measure of angle x.

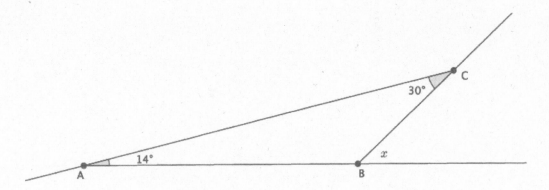

Example 2

Find the measure of angle x.

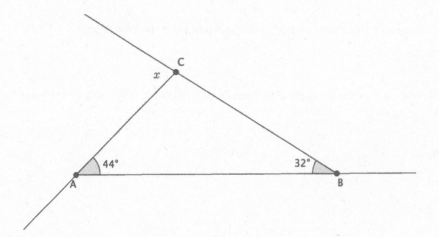

EUREKA
MATH®

Example 3

Find the measure of angle x.

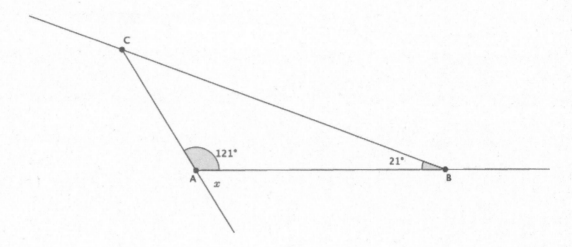

Example 4

Find the measure of angle x.

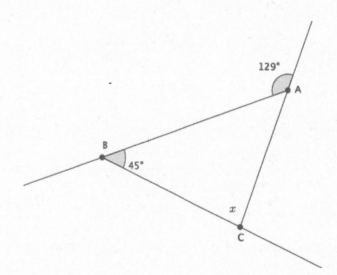

Exercises 5–10

5. Find the measure of angle x. Present an informal argument showing that your answer is correct.

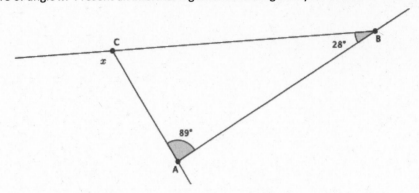

6. Find the measure of angle x. Present an informal argument showing that your answer is correct.

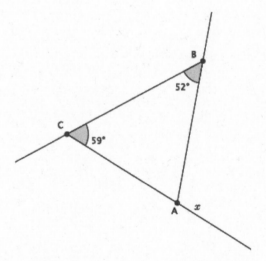

EUREKA
MATH

7. Find the measure of angle x. Present an informal argument showing that your answer is correct.

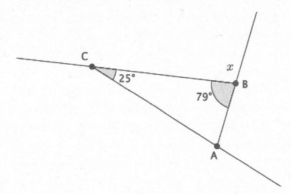

8. Find the measure of angle x. Present an informal argument showing that your answer is correct.

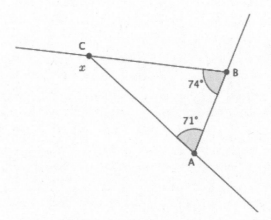

9. Find the measure of angle x. Present an informal argument showing that your answer is correct.

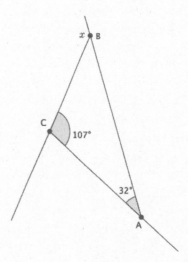

EUREKA
MATH®

10. Find the measure of angle x. Present an informal argument showing that your answer is correct.

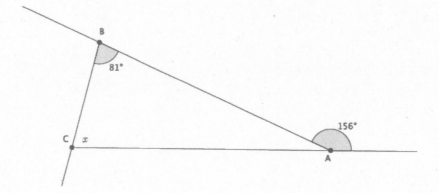

Lesson 14:   More on the Angles of a Triangle

EUREKA
MATH

Lesson Summary

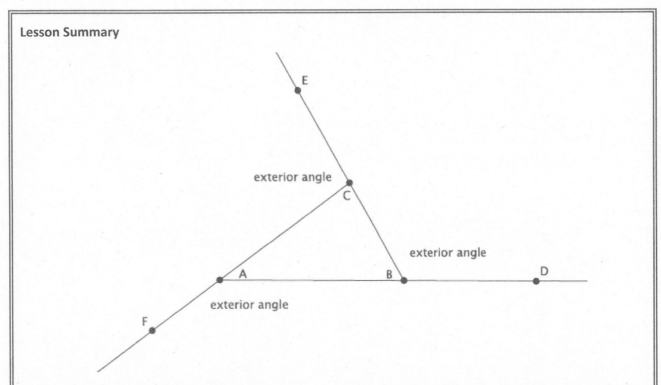

The sum of the measures of the remote interior angles of a triangle is equal to the measure of the related exterior angle. For example, $\angle CAB + \angle ABC = \angle ACE$.

Name _____   Date _____

1. Find the measure of angle p. Present an informal argument showing that your answer is correct.

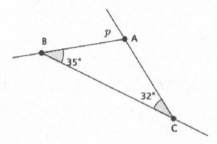

2. Find the measure of angle q. Present an informal argument showing that your answer is correct.

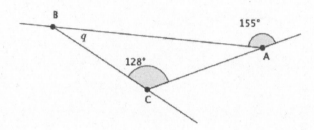

3. Find the measure of angle r. Present an informal argument showing that your answer is correct.

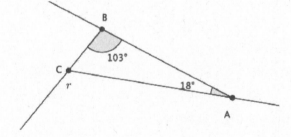

1. Find the measure of angle x. Present an informal argument showing that your answer is correct.

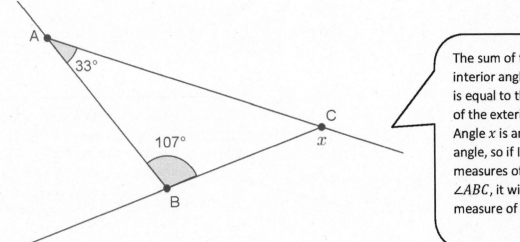

> The sum of the remote interior angle measures is equal to the measure of the exterior angle. Angle x is an exterior angle, so if I add the measures of $\angle CAB$ and $\angle ABC$, it will equal the measure of angle x.

Since $33 + 107 = 140$, the measure of angle x is $140°$. We know that triangles have a sum of interior angle measures that is equal to $180°$. We also know that straight angles measure $180°$. The measure of $\angle ACB$ must be $40°$, which means that the measure of $\angle x$ is $140°$.

> There is a straight angle comprised of $\angle ACB$ and $\angle x$ that will equal $180°$.

2. Write an equation that would allow you to find the measure of ∠y. Present an informal argument showing that your answer is correct.

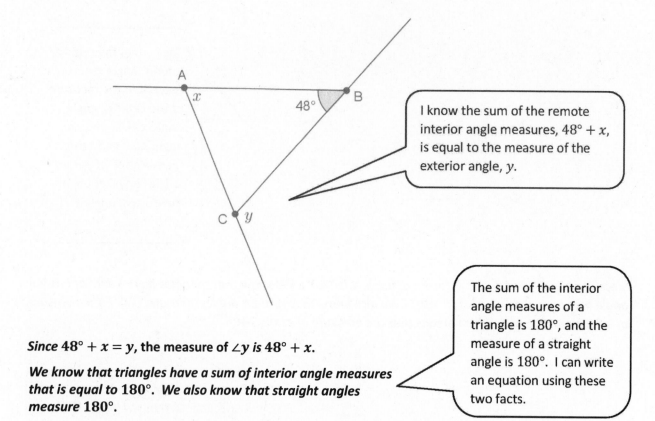

I know the sum of the remote interior angle measures, $48° + x$, is equal to the measure of the exterior angle, y.

The sum of the interior angle measures of a triangle is $180°$, and the measure of a straight angle is $180°$. I can write an equation using these two facts.

Since $48° + x = y$, the measure of ∠y is $48° + x$.

We know that triangles have a sum of interior angle measures that is equal to $180°$. We also know that straight angles measure $180°$.

The measure of the interior angles of the triangle is $x + 48° + ∠ACB$.

The measure of the straight angle is $y + ∠ACB$.

Then, $x + 48° + ∠ACB = 180°$, and $y + ∠ACB = 180°$. Since both equations are equal to $180°$, then $x + 48° + ∠ACB = y + ∠ACB$. Subtracting ∠ACB from each side of the equation yields $x + 48° = y$.

EUREKA MATH

For each of the problems below, use the diagram to find the missing angle measure. Show your work.

1. Find the measure of angle x. Present an informal argument showing that your answer is correct.

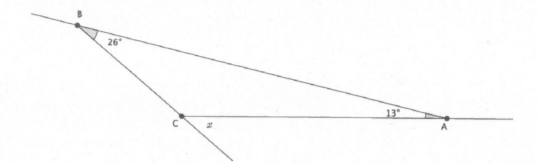

2. Find the measure of angle x.

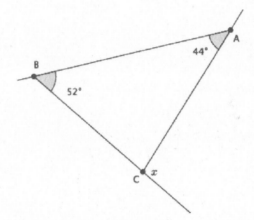

3. Find the measure of angle x. Present an informal argument showing that your answer is correct.

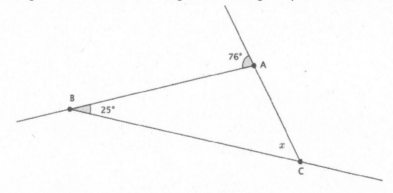

EUREKA
MATH

4. Find the measure of angle x.

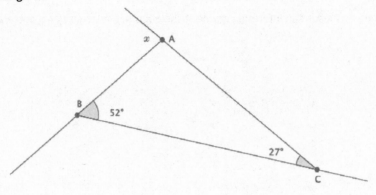

5. Find the measure of angle x.

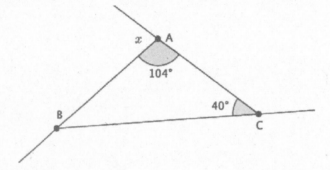

6. Find the measure of angle x.

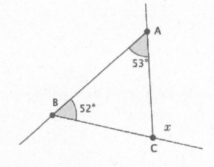

EUREKA
MATH

7. Find the measure of angle x.

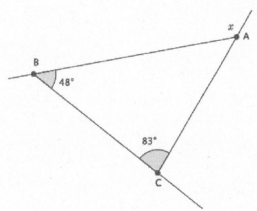

8. Find the measure of angle x.

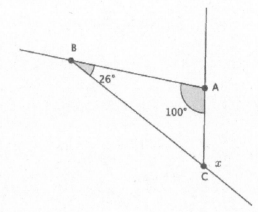

9. Find the measure of angle x.

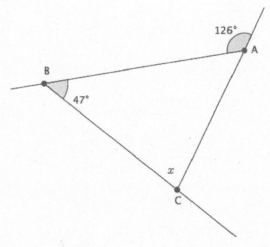

10. Write an equation that would allow you to find the measure of angle x. present an informal argument showing that your answer is correct.

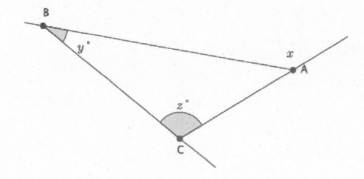

EUREKA
MATH

Example 1

Now that we know what the Pythagorean theorem is, let's practice using it to find the length of a hypotenuse of a right triangle.

Determine the length of the hypotenuse of the right triangle.

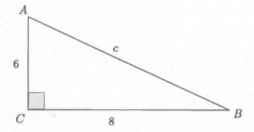

The Pythagorean theorem states that for right triangles $a^2 + b^2 = c^2$, where a and b are the legs, and c is the hypotenuse. Then,

$$a^2 + b^2 = c^2$$
$$6^2 + 8^2 = c^2$$
$$36 + 64 = c^2$$
$$100 = c^2.$$

Since we know that $100 = 10^2$, we can say that the hypotenuse c is 10.

Example 2

Determine the length of the hypotenuse of the right triangle.

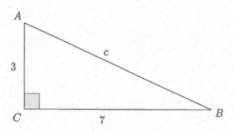

Exercises 1–5

For each of the exercises, determine the length of the hypotenuse of the right triangle shown. Note: Figures are not drawn to scale.

1.

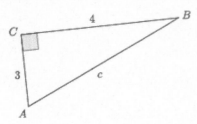

2.

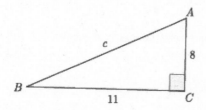

3.

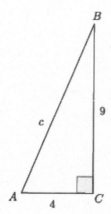

EUREKA
MATH®

4.

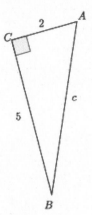

5.

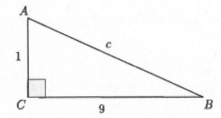

Lesson Summary

Given a right triangle ABC with C being the vertex of the right angle, then the sides $\overline{AC}$ and $\overline{BC}$ are called the *legs* of $\triangle ABC$, and $\overline{AB}$ is called the *hypotenuse* of $\triangle ABC$.

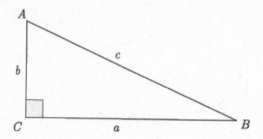

Take note of the fact that side a is opposite the angle A, side b is opposite the angle B, and side c is opposite the angle C.

The Pythagorean theorem states that for any right triangle, $a^2 + b^2 = c^2$.

Name _____   Date _____

1. Label the sides of the right triangle with leg, leg, and hypotenuse.

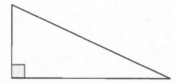

2. Determine the length of c in the triangle shown.

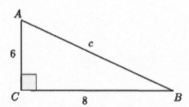

3. Determine the length of c in the triangle shown.

For each of the problems below, determine the length of the hypotenuse of the right triangle shown. Note: Figures are not drawn to scale.

1.

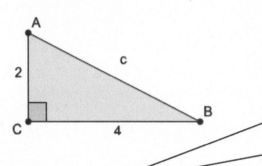

$$a^2 + b^2 = c^2$$
$$2^2 + 4^2 = c^2$$
$$4 + 16 = c^2$$
$$20 = c^2$$

I know that 2 and 4 are the legs of the triangle. I know this because the hypotenuse is across from the 90° angle. Since the hypotenuse is side c in my formula, I substitute the 2 and 4 for a and b.

Since I do not know what number times itself produces 20, for now I can leave my answer as $20 = c^2$.

2.

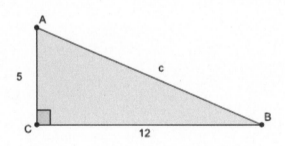

$$a^2 + b^2 = c^2$$
$$12^2 + 5^2 = c^2$$
$$144 + 25 = c^2$$
$$169 = c^2$$
$$13 = c$$

Since I know that $13 \times 13 = 169$, then I know that $c = 13$.

EUREKA MATH

For each of the problems below, determine the length of the hypotenuse of the right triangle shown. Note: Figures are not drawn to scale.

1.

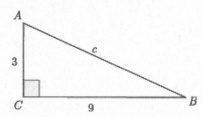

2.

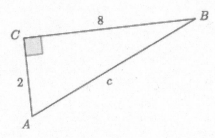

3.

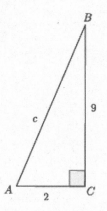

4.

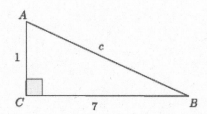

5.

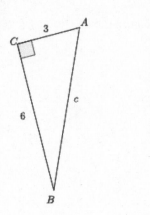

6.

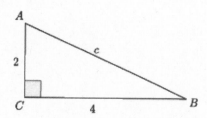

7.

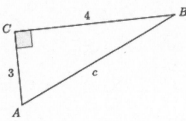

8.

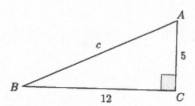

9.

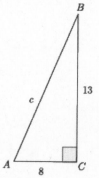

EUREKA
MATH

10.

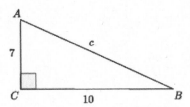

11.

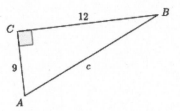

12.

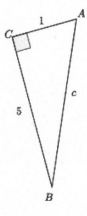

EUREKA
MATH

Example 1

Given a right triangle with a hypotenuse with length 13 units and a leg with length 5 units, as shown, determine the length of the other leg.

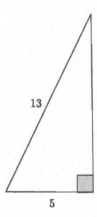

$$5^2 + b^2 = 13^2$$
$$5^2 - 5^2 + b^2 = 13^2 - 5^2$$
$$b^2 = 13^2 - 5^2$$
$$b^2 = 169 - 25$$
$$b^2 = 144$$
$$b = 12$$

The length of the leg is 12 units.

Exercises 1–2

1. Use the Pythagorean theorem to find the missing length of the leg in the right triangle.

EUREKA MATH

2. You have a 15-foot ladder and need to reach exactly 9 feet up the wall. How far away from the wall should you place the ladder so that you can reach your desired location?

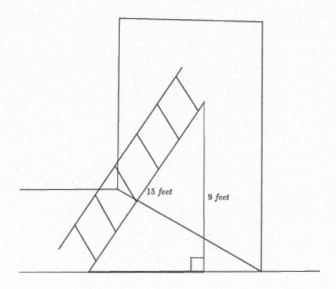

Exercises 3–6

3. Find the length of the segment AB, if possible.

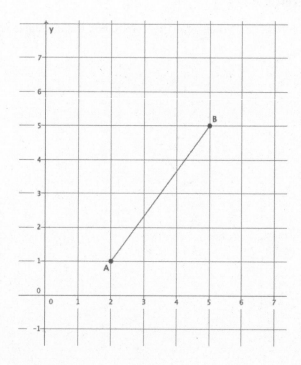

Lesson 16: Applications of the Pythagorean Theorem

EUREKA MATH®

4. Given a rectangle with dimensions 5 cm and 10 cm, as shown, find the length of the diagonal, if possible.

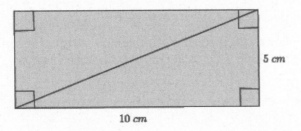

5 cm

10 cm

5. A right triangle has a hypotenuse of length 13 in. and a leg with length 4 in. What is the length of the other leg?

6. Find the length of b in the right triangle below, if possible.

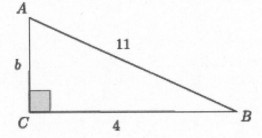

Lesson Summary

The Pythagorean theorem can be used to find the unknown length of a leg of a right triangle.

An application of the Pythagorean theorem allows you to calculate the length of a diagonal of a rectangle, the distance between two points on the coordinate plane, and the height that a ladder can reach as it leans against a wall.

Lesson 16:   Applications of the Pythagorean Theorem

EUREKA MATH

Name _____   Date _____

1. Find the length of the missing side of the rectangle shown below, if possible.

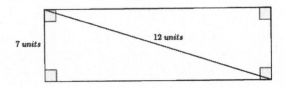

2. Find the length of all three sides of the right triangle shown below, if possible.

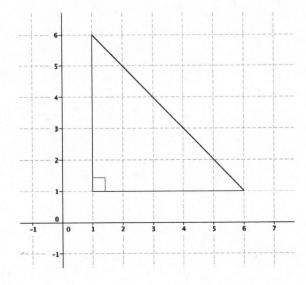

1. Find the length of the segment AB shown below, if possible.

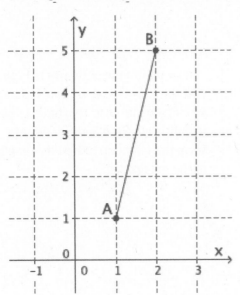

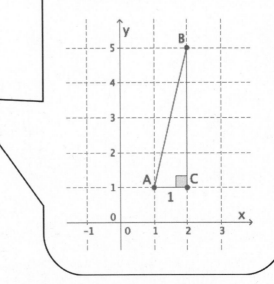

I know that the grid lines on the coordinate plane meet at a right angle. I can make my right triangle using a horizontal line through point A and a vertical line through point B.

$$a^2 + b^2 = c^2$$
$$1^2 + 4^2 = c^2$$
$$1 + 16 = c^2$$
$$17 = c^2$$

2. A rectangle has dimensions 6 cm by 8 cm. What is the length of the diagonal of the rectangle?

$$a^2 + b^2 = c^2$$
$$6^2 + 8^2 = c^2$$
$$36 + 64 = c^2$$
$$100 = c^2$$
$$10 = c$$

The length of the diagonal is **10** cm.

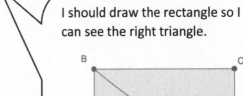

I should draw the rectangle so I can see the right triangle.

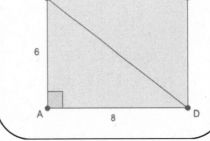

3. Determine the length of the unknown side, if possible.

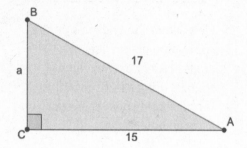

I know that the hypotenuse is 17 and that the hypotenuse is represented by c in my formula. This time, I need to substitute for b and c and then solve the equation to find the length of the missing leg.

$$a^2 + b^2 = c^2$$
$$a^2 + 15^2 = 17^2$$
$$a^2 + 225 = 289$$
$$a^2 + 225 - 225 = 289 - 225$$
$$a^2 = 64$$
$$a = 8$$

Lesson 16:   Applications of the Pythagorean Theorem

EUREKA
MATH

1. Find the length of the segment AB shown below, if possible.

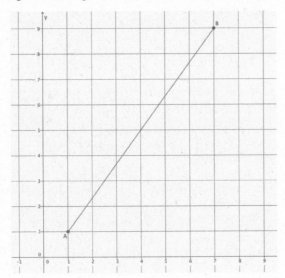

2. A 20-foot ladder is placed 12 feet from the wall, as shown. How high up the wall will the ladder reach?

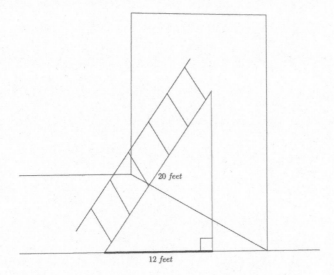

3. A rectangle has dimensions 6 in. by 12 in. What is the length of the diagonal of the rectangle?

Use the Pythagorean theorem to find the missing side lengths for the triangles shown in Problems 4–8.

4. Determine the length of the missing side, if possible.

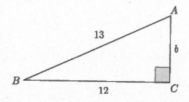

5. Determine the length of the missing side, if possible.

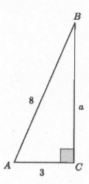

6. Determine the length of the missing side, if possible.

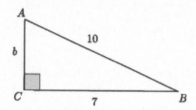

7. Determine the length of the missing side, if possible.

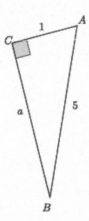

Lesson 16:   Applications of the Pythagorean Theorem

EUREKA
MATH

8. Determine the length of the missing side, if possible.

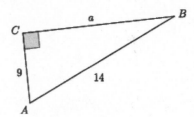

Credits

Great Minds® has made every effort to obtain permission for the reprinting of all copyrighted material. If any owner of copyrighted material is not acknowledged herein, please contact Great Minds for proper acknowledgment in all future editions and reprints of this module.